Peter

With best wishes

Robin Jones

13/1/93

Modern Coloproctology

Modern Coloproctology

Surgical Grand Rounds From St Mark's Hospital

Edited by

Robin Phillips
and
John Northover

Edward Arnold
A division of Hodder & Stoughton
LONDON MELBOURNE AUCKLAND

First published in Great Britain 1993

Full CIP data is available from the British Library

ISBN 0 340 55258 1

Typeset in 10/11pt Linotron Palatino by Rowland Phototypesetting Limited, Bury St Edmunds, Suffolk.
Printed and bound in Great Britain for Edward Arnold, a division of Hodder and Stoughton Limited, Mill Road, Dunton Green, Sevenoaks, Kent TN13 2YA by Butler and Tanner Limited, Frome, Somerset.

Preface

These are exciting times in coloproctology, as technical innovation in management and our understanding of the aetiology of several colorectal diseases are moving forward rapidly. At St Mark's one of the ways we try to keep abreast of developments is our weekly Friday morning Grand Rounds, at which hospital staff and our visitors from around the world present for discussion areas of current interest and controversy. The topics vary from the extremely practical to the theoretically exciting – our only criterion for choosing topics is that they should be of current and general interest to those within our specialty.

The topics covered in this first selection of papers taken from our meetings reflect the diversity of subjects that we like to discusss. We hope that they pass on the stimulation that we have felt as participants in St Mark's Hospital's Grand Rounds over the past year.

Robin Phillips,
John Northover,
St Mark's Hospital,
1992

Contents

List of Contributors

Clive I Bartram FRCP, FRCR, DMRD
Consultant Radiologist, St Mark's Hospital, London.

Sarah J D Burnett MRCP
Senior Registrar, St Bartholomew's Hospital, London.

Michael J G Farthing BSc, MD, FRCP
Professor Gastroenterology, St Bartholomew's Hospital, London; Honorary Consultant Gastroenterologist, St Luke's Hospital for the Clergy; Honorary Consultant in Gastroenterology to the Army.

Sven Goldman MD, PhD
Department of Surgery, Sodersjukhuset, Stockholm, Sweden.

Thomas Ihre, MD, PhD
Department of Surgery, Sodersjukhuset, Stockholm, Sweden.

Michael Levitt FRACS
Consultant Surgeon, Colorectal Surgical Service, Sir Charles Gairdner Hospital, Nedlands, Australia.

Peter J Lunniss BSc, FRCS
Clinical Lecturer in Surgery, St Bartholomew's Hospital; Research Registrar, St Mark's Hospital, London.

Jane L McCue MBBS, FRCS
Research Registrar, Professorial Surgical Unit, St Bartholomew's Hospital, London.

R John Nicholls M Chir, FRCS
Consultant Surgeon, St Mark's Hospital; St Thomas' Hospital, London.

John M A Northover MS, FRCS
Consultant Surgeon, St Mark's Hospital; Director, ICRF Colorectal Cancer Unit; Senior Lecturer, St Bartholomew's Hospital Medical College, London.

Robin K S Phillips MS, FRCS
Consultant Surgeon, St Mark's Hospital; The Homerton Hospital, London.

David A Rothenberger MD, FACS
Clinical Professor of Surgery; Chief, Division of Colon and Rectal Surgery, University of Minnesota, Minneapolis, USA.

Adam D N Scott BSc, MS, FRCS
Senior Surgical Registrar, St Bartholomew's Hospital; Resident Surgical Officer, St Mark's Hospital, London.

Asha Senapati PhD, FRCS
Consultant Surgeon, Mayday University Hospital, Croydon.

James P S Thomson MS, FRCS, DObst, RCOG
Consultant Surgeon and Clinical Director, St Mark's Hospital, London.

Norman Waterhouse FRCS, FRCS (Plastic Surgery)
Consultant Plastic Surgeon, Charing Cross Hospital; St Mary's Hospital; Westminster Hospital, London.

J Graham Williams MCh, FRCS
Lecturer in Surgery, Department of Surgery, University of Birmingham, Birmingham.

Julia E C W Verne MBBS
ICRF Unit, St Mark's Hospital, London.

Foreword

St Mark's Hospital has been the beacon in the night for colon and rectal surgery for over 150 years. As a recipient of the learning experience at St Mark's, I am honoured to have the opportunity to review this text prior to publication and particularly pleased to have the opportunity to write this brief comment. This book is special because it focuses on the controversial aspects of this field of surgery in 1992. The authors of each chapter have succeeded in presenting an in-depth analysis of the clinical problem. Undoubtedly, surgeons will find the book valuable in the day-to-day management of their patients.

We need only to scan this volume to realize how times have changed and how the field has expanded. The book also amply shows us that new information is accumulating rapidly, and that we still have much to learn. From research that is currently in progress, we can anticipate new approaches as well as improved technologies for the management of our patients.

One of the outstanding features of St Mark's Hospital for many years has been its teaching courses for young surgeons. The chapters in this book are an enlargement of some of these lectures with pertinent and up-to-date references.

Through much of this century the surgeons, pathologists and radiologists of St Mark's Hospital have assumed a leadership role of an international nature. This book is just another example of their continued pre-eminence in the field of coloproctology. We are all indebted to the Editors for their work and effort to make this text possible.

Stanley M. Goldberg, M.D., F.A.C.S., F.R.A.C.S.(Hon.)
Clinical Professor of Surgery
Division of Colon and Rectal Surgery
Department of Surgery
University of Minnesota Medical School
1992

1

Colectomy with ileorectal anastomosis or restorative proctocolectomy for familial adenomatous polyposis

R. John Nicholls

Introduction

With the establishment of the polyposis register at St Mark's Hospital in the 1920s, the concept of prophylactic colectomy became well established. Colectomy and ileorectal anastomosis (IRA) in the treatment of familial adenomatous polyposis (FAP) was introduced in 1948. It was adopted as the favoured operation for this condition over total proctocolectomy with the sole aim of avoiding an ileostomy. This procedure, which avoided an ileostomy in patients who were often in their teens and could be justified on pathological grounds, had much to recommend it.

The price paid was a certain incidence of cancer in the rectal stump.[1] Moertel *et al.*[2] observed 143 patients having a colonic resection at the Mayo Clinic and reported a rising incidence of rectal cancer as the length of follow-up increased; beyond 20 years this was over 50 per cent. The risk appeared to increase with greater numbers of rectal adenomas, and also if a carcinoma already existed in the colon. In an updated report of the same 143 patients (all of whom had adenomas in the rectum), Bess *et al.*[3] reported that 46 (32 per cent) had developed rectal cancer during a median follow-up period of 19.1 years. They projected the estimate of rectal cancer at 30 years to be 55 per cent. They also showed a cumulative incidence of rectal cancer with the length of interval from colectomy.

Others studies have shown similar results. For example, Watne *et al.*[4] reported rectal cancer in 7 (22 per cent) of a series of 32 patients followed for a mean of 14 years after colectomy with IRA.

These rather high incidences might be explained by the length of follow-up (as seen in the Mayo Clinic series) but other factors could be influential. Regular surveillance of the rectal stump was not possible in all patients and in some cases the operation left a significant length of sigmoid colon. Furthermore, there may be epidemiological reasons for a higher incidence in one country compared with another.

At St Mark's Hospital the experience was somewhat different. Between

1948 and 1984, 174 patients underwent colectomy with ileorectal anastomosis. The effectiveness of the policy of prophylaxis was demonstrated by an incidence of cancer already present in 20.3 per cent of patients presenting with symptoms compared with a rate of 5.2 per cent among those relatives of known affected patients who were called up as a result of the screening programme. (Overall 10.3 per cent of the whole group of 174 patients had a cancer at presentation). During the period of follow-up (median 12 years, range 1–41 years) 11 patients developed a cancer in the rectum. Of these only three died. Overall, 35 patients in the whole series died with 21 related to the disease or its treatments. The three cases of rectal cancer compare with six deaths from upper gastrointestinal malignancy and three deaths due to desmoid disease. Although a weakness of the procedure, the development of rectal cancer was less than in the American series. Thus at 20 years, the cumulative risk was around 10 per cent (Fig. 1.1). Perhaps this was because of the management policy of sigmoidoscopic examination of the rectal stump at 3–6 monthly intervals. It may be significant that two of the three deaths were among 'poor' attenders and also that four (37 per cent) of the 11 cases developing carcinoma were of Dukes A stage.[5]

Nevertheless, there is a significant incidence of rectal cancer after colectomy with ileorectal anastomosis, with a clear tendency for this to rise with

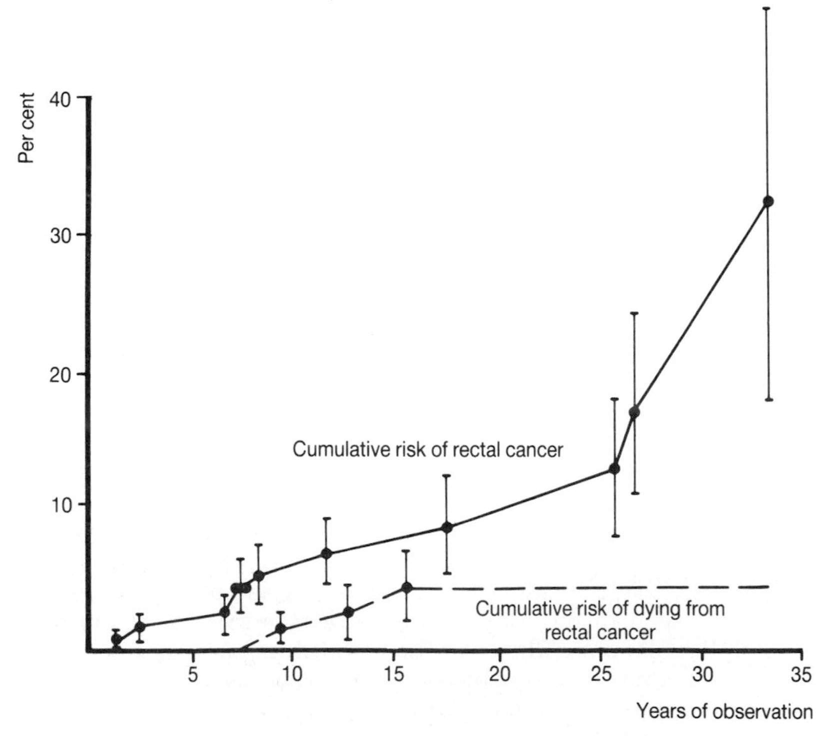

Fig. 1.1 Cumulative risk of rectal cancer after colectomy with ileorectal anastomosis.

the length of follow-up. The attraction of restorative proctocolectomy is obvious. In theory it should abolish any development of large bowel cancer.

Ileorectal anastomosis versus restorative proctocolectomy

Before jumping to the conclusion that FAP should be treated by a pouch procedure, it should be compared with colectomy and ileorectal anastomosis in other ways including morbidity, mortality and function. As indicated above, there is a further consideration in this debate. With increasing knowledge of the natural history of the disease, it has become apparent that rectal cancer after colectomy with IRA is by no means the most common cause of death. As shown in Table 1.1 upper gastrointestinal malignancy and desmoid disease accounted for six and three deaths respectively compared with those from rectal stump carcinoma. Thus, the theoretical avoidance of rectal cancer by restorative proctocolectomy will only make a small albeit important contribution to the reduction in mortality.

Table 1.1 Familial adenomatous polyposis – related deaths following colectomy with ileorectal anastomosis (n = 174)

Carcinoma	
Colon	4
Rectal stump	3
Upper gastrointestinal tract	6
Unknown site	2
Other causes	
Desmoid tumour	3
Small bowel obstruction	1
Other	2
Operative	1

Reproduced by permission of the publishers, Butterworth-Heinemann Ltd, from Bussey *et al.*[5]

The study

Aims

We set out to compare the mortality, morbidity and function by an analysis of the short-term outcome of each operation carried out at St Mark's Hospital during the same period. This started in July 1977 with the first restorative proctectomy for FAP. Between July 1977 and December 1989, 82 patients underwent a colectomy with IRA and 38 an ileoanal reservoir procedure. Of the former 20 were not studied; six had died, three had emigrated and in 11 it was not possible to obtain data on function. Thus 62 cases were studied. Of these, none had a carcinoma in the large bowel.

Of the 38 patients who underwent an ileoanal reservoir procedure one had emigrated leaving 37 for study. 13 of these cases had had a previous large bowel resection including colectomy with IRA in 11 and a straight ileoanal reconstruction in two. In these 13 patients the indications included uncontrollable rectal adenomas (seven), rectal cancer (three) and the age of the patient (three). In the remaining 24 patients the indications included uncontrollable rectal polyps (three), rectal cancer (four), colonic cancer (five), and the age of the patient (12).

Information on morbidity was obtained from the patients' hospital records. With regard to function, a personal assessment was made wherever possible. A standard proforma was used and the patient was interviewed by a member of the staff of the Polyposis Registry. In this way it was hoped to avoid surgeon bias.

Basic details are shown in Table 1.2. It can be seen that a significant difference in the mean age in the ileoanal reservoir and IRA groups existed. This is explained by case selection whereby patients diagnosed in adolescence as a result of family screening tended to have an IRA whereas the older age group presenting for the first time with symptomatic disease tended to have an ileoanal procedure. Thus the groups were not strictly comparable, a consideration that might have relevance to function (see below).

The Chi squared test with Yates' correction was used to compare proportions for large samples, and Fisher's exact test for small samples. Normally distributed data were compared by Student's t test and non-parametric using Wilcoxon's signed rank test. A probability value of less than 5 per cent was taken to be significant. The tables clearly show the denominator of each proportion calculated, because not all data were obtained from every patient.

Results

There was no postoperative mortality. Morbidity was assessed using the following parameters; length of hospital stay, incidence of postoperative

Table 1.2 Restorative proctocolectomy (RPC) and colectomy with ileorectal anastomosis (IRA) 1977–89

	RPC	IRA
Patients studied	37	62
male/female	18/19	33/29
Age (years)		
median	31*	19*
range	14–52	12–53
Follow-up (years)		
mean	5.2	6.5
range	0.7–12.2	0.7–12.8

*$p < 0.001$.

complications, incidence of re-laparotomy for complications and the time taken for return to normal activity (Table 1.3). All patients in the ileoanal group had a temporary ileostomy that was subsequently closed in all but three cases.

The duration of hospital stay was significantly greater after an ileoanal procedure including stoma closure. It is of interest to note, however, that when the admission for stoma closure was excluded, the median values of 15 and 11 days for in-patient stay following the ileoanal and ileorectal anastomosis respectively were considerably less different. Furthermore, the duration of hospital stay for ileoanal patients who had had a previous colectomy was 12 compared with 17 days for those having a restorative proctocolectomy as a first procedure.

The complication rate was markedly different. It is noteworthy that complications due to the ileostomy accounted for a third of all complications (seven of 21). Moreover, three out of the eight re-laparotomies were for ileostomy-related complications, two of which were intestinal obstruction.

There was a significant difference in the time taken to resume normal activity. This was mainly owing to the two-stage management of the ileoanal reservoir procedure.

As far as specific complications are concerned (Table 1.4) the two most important were intestinal obstruction and pelvic sepsis with or without an evident defect of the anastomosis. Obstruction occurred in only one case after the ileoanal operation itself, but six more cases followed closure of the ileostomy. Of these seven, three required a laparotomy, a rate of 8 per cent. In contrast, none of the seven patients having ileorectal anastomosis who developed obstruction required a laparotomy. Pelvic sepsis or an anastomotic defect occurred in six (17 per cent) of the patients. Whereas this may seem a high incidence, the three cases with a defect healed without incident and did not lead to any delay of closure of the ileostomy. Thus three (8 per cent) of the 36 patients had significant sepsis in the

Table 1.3 General parameters of morbidity in the two groups

	RPC	RPC + stoma closure	IRA
Patients reviewed	36	33	62
Hospital stay (days)			
median	15	24*	11*
range	9–43	17–63	8–44
Complications	14 (38)	21 (57)	(21)**
Re-laparotomy	5 (14)	8 (22)	2 (3)†
Time to normal activity (weeks)			
median		30	14
range		4–98	2–52

*$p < 0.001$; **$p < 0.0002$; †$p < 0.01$; RPC, restorative proctocolectomy; IRA, ileorectal anastomosis; percentages in brackets.

Table 1.4 Specific complications

	RPC	RPC + stoma closure	IRA
Patients reviewed	36	33	62
Intestinal obstruction	1 (1)	6 (2)	7
Anastomotic defect	3		1 (1)
Pelvic abscess	3		1
Intra-abdominal haemorrhage	2 (2)		
Anastomotic stricture	1 (1)		1 (1)
Wound infection			2
Deep venous thrombosis	1		1
Stoma retraction	1 (1)		
Repeat mucosectomy		1 (1)	

Operation number in brackets; RPC, restorative proctocolectomy; IRA, ileorectal anastomosis.

pelvis. None of these patients, however, required further surgery for this complication.

We also looked at the number of outpatient visits required per annum in the two groups. After ileoanal anastomosis, there was a mean of 2.8/year compared with 2.1/year after IRA. There was, however, a difference in the number of inpatient admissions. 27 (44 per cent) of patients after IRA required a mean of 2.5 admissions to hospital, each of about 3 days duration, for fulguration of rectal adenomas over a mean follow-up period of 6.5 years. None in the ileoanal group required re-admission.

There were, therefore, 202 patient days required for inpatient treatment after IRA of rectal adenomas equivalent to 117 patient days for a population of 36 patients. This figure is considerably less than the inpatient stay required for closure of the ileostomy in the ileoanal patients, which amounted to about 430 patient days based on a median inpatient stay for stoma closure of 12 days.

Function

Details of function were obtained from 30 of the 36 patients having an ileoanal anastomosis and in 61 after IRA. The results are shown in Table 1.5. Frequency is recorded both pre- and postoperatively. It can be seen that patients having an ileoanal procedure had a higher frequency than those having IRA before any surgery was carried out. The former group was older by about 10 years, but age should not be an important factor in this respect. It could well be that a proportion of patients selected for an ileoanal procedure were already symptomatic from their disease. In addition, those who had already had an ileorectal anastomosis would be expected to have a higher frequency than unoperated unsymptomatic patients.

Table 1.5 Function

	RPC	IRA	p
Patients reviewed	33	62	
Frequency defaecation			
/24 hour preoperation	5.2 (1–20)	2 (0.5–20)	< .001
postoperation median (range)	4.4 (1.5–8)	3 (1–11)	< .001
Night evacuation			
(> × 1 week)	13/30 (43)	6/61 (10)	< .001
Urgency < 15 minutes	5/29 (17)	31/62 (50)	< .01
Antidiarrhoeal medication	9/29	10.60 (17)	
Continence			
normal	18/30	43/60 (72)	
pad	3/30 (10)	0/24	
Anal irritation	23/30	37/60 (62)	
Reduced quality of life			
work	5/28	3/60 (5)	
social life	5/28	9/60 (15)	

Percentages in brackets.

When the functional results after each operation were compared, it was found that frequency following ileoanal anastomosis fell slightly whereas it rose after ileorectal anastomosis. The proportion of patients needing to defaecate at night more than once a week was, however, less after ileorectal anastomosis (10 per cent) than after ileoanal anastomosis (40–45 per cent). In other respects (continence, anal irritation, dietary difficulties, anti-diarrhoeal medication, quality of life) there was no significant difference. In the case of urgency, ileoanal scored better than ileorectal anastomosis.

Discussion

The two groups were different in certain respects. Clinically those having IRA were mostly adolescents who had been identified through family screening to have the disease. In contrast, the ileoanal group contained many who had already developed symptoms. This accounts for the age difference, the difference in prevalence of large bowel cancer at presentation and the proportion of patients who had already had a colectomy. Whereas these factors might *per se* have accounted to some extent for the differences in results, it is nevertheless abundantly clear that the ileoanal procedure in our hands led to greater morbidity in whichever way assessed. To some degree, the temporary ileostomy was responsible for much of this morbidity. Therefore it ought to be the case that a one stage restorative proctocolectomy would be associated with fewer complications. It would certainly reduce the treatment time and allow patients to return to normal activity earlier. Everett and Pollard[6] have shown that omitting an ileostomy does not result in a higher morbidity of the procedure itself. Others have also promulgated the avoidance of a temporary ileostomy.[7] There are, however, disadvantages; firstly when a severe complication such as ileoanal dehiscence with pelvic sepsis does occur, an emergency re-laparotomy may be necessary. This occurred in three out of the 16 cases reported by Kmiot and Keighley.[7] Whereas this may be an acceptable risk, the patients in whom such action is necessary are severely ill. This might not have been the case were a pre-existing ileostomy to have been in place. The second consideration is subtle. Function after restorative proctocolectomy is not as good in most polyposis patients as it was preoperatively. (The same can be said for colectomy with IRA). Therefore it might occur that a patient could be dissatisfied with the postoperative function without being able to compare his or her state with life with an ileostomy. If an ileostomy is the alternative to a restorative proctocolectomy, it certainly helps in long-term postoperative management to have a patient who knows what an ileostomy is like.

With regard to function, there is little to choose between the two operations. Frequency of defaecation compared with preoperative function is not particularly different and continence and urgency are balanced between the two. Logistically the number of outpatient visits was similar after each operation. Fulguration of rectal polyps after ileorectal anastomosis might be performed in future as an outpatient procedure.

The choice should be based upon the perceived risk of cancer developing in the rectum. To date certain factors exist that might influence this. Age of presentation is the most important. The significant difference in the presence of cancer in adolescents compared with patients over 30 years of age should influence management. In a young patient the risk of rectal cancer either being present or occurring within 10 years is very small. Here a colectomy with IRA should be the treatment of choice. For those presenting with symptoms with a considerably increased risk of cancer already being present, the indication for a restorative proctocolectomy is strong. Whereas this policy can be justified by the likelihood of cancer already having developed at presentation, there is less support for it on the basis of an increased cancer risk with respect to time following colectomy with IRA. In following such patients, we have found at St Mark's that cancer in the rectal stump occurs somewhat haphazardly and not clearly as a function of the duration of the interval from the original operation. The Mayo Clinic experience suggests, however, that it may be,[3] and the clinician must be guided by the vital aim of preventing the occurrence of rectal cancer. Whether precisely quantifiable or not, the risk is greater with age. Other indications for a restorative proctocolectomy must include patients who have uncontrollable rectal adenomas and those who already have developed a large bowel carcinoma. In the latter group evidence exists that the risk of rectal cancer after colectomy with ileorectal anastomosis is increased by a factor of four.[2,8] In the former group, large numbers of rectal adenomas at the time of presentation do not necessarily indicate a restorative proctocolectomy. Evidence exists that rectal polyps can regress after colectomy with ileorectal anastomosis.[9] It is probably preferable, therefore, to advise an ileorectal anastomosis in a young patient even with many rectal adenomas, in the hope that regression will follow.

An essential requirement for colectomy with ileorectal anastomosis is the availability of the patient for follow-up. If this cannot be relied upon, then a total removal of all large bowel mucosa is preferable. This would constitute an indication for restorative proctocolectomy.

References

1 Beart R W. Familial polyposis. *Br J Surg*. 1985; **72** (suppl): S31–2.
2 Moertel C G, Hill J R, Adson M A. Surgical management of multiple polyposis. *Arch Surg*. 1971; **100**: 521–6.
3 Bess M A, Adson M A, Moertel C G. Rectal cancer following colectomy for polyposis. *Arch Surg*. 1980; **115**: 460–67.
4 Watne A L, Carrier J M, Durham J P *et al.*. The occurrence of carcinoma of the rectum following ileoproctostomy for familial polyposis. *Ann Surg*. 1983; **197**: 550–54.
5 Bussey H J R, Eyers A A, Ritchie S M, Thomson J P S. The rectum in adenomatous polyposis: the St Mark's policy. *Br J Surg*. 1985; **72** (suppl): S29–30.
6 Everett W G, Pollard S G. Restorative proctocolectomy without temporary ileostomy. *Br J Surg*. 1990; **77**: 621–2.
7 Kmiot W A, Keighley M R B. Totally stapled abdominal restorative proctocolectomy. *Br J Surg*. 1989; **76**: 961–4.
8 Bussey H J R. Personal communication, 1990.
9 Nicholls R J, Springall R G, Gallagher P. Regression of rectal adenomas after colectomy and ileorectal anastomosis for familial adenomatous polyposis. *Br Med J*. 1988; **246**: 1707–708.

2

Reconstructive surgery of the groin and perineum

Norman Waterhouse and John M.A. Northover

The role of reconstructive plastic surgery is to achieve wound healing, preserve or restore function and provide the best possible cosmesis. Wounds and defects in the groin and perineum frequently present special problems of management. These are due to a combination of their anatomical position, propensity for infection and the underlying aetiology. Thus proximity to the anus, faecal soiling, heat, moisture and intertrigo present obvious problems. Inevitable mobility and the difficulty of applying fixed dressings complicate management of these wounds. Resident flora increase the likelihood of contamination of grafts, flaps and suture lines. Patients undergoing extensive perineal surgery are often elderly and/or malnourished. Sacral, ischial and trochanteric pressure sores, extensive hidradenitis suppurativa and cancer surgery illustrate these problems.

Perhaps the most difficult reconstructive challenge is the perineal wound that follows total or partial pelvic exenteration. After abdominoperineal excision, healing may occur slowly,[1] particularly in cases of Crohn's disease or after radiotherapy. If wound breakdown occurs, secondary healing may take months involving repeated attendance, nursing care, dressings and a protracted hospital stay. The presence of a large cavity allows loops of bowel to settle in the pelvis. This predisposes to bowel obstruction and hernia formation and prevents postoperative radiotherapy to the pelvic field. Although most pelvic exenterations are performed for malignant disease, extensive perineal fistulae and radiation damage may also require radical excision.

Despite the acceptance that many of these wounds take time to heal, there are many reasons why early rapid primary healing should be achieved. Part of the postoperative metabolic disturbances affecting these patients is due to the 'pelvic burn'. Following a pelvic exenteration the raw surface may equate to an area of 20 per cent skin loss from a burn. Early obliteration of this defect removes this problem. Further, the mechanical effect of obliterating the pelvic dead space is to prevent gravitation of the bowel into the pelvis, leading to herniation and obstruction, and allows administration of postoperative radiotherapy. Primary reconstruction of the perineum may also include vaginal reconstruction and treatment of vesico-vaginal fistula thus allowing women the possibility of early sexual rehabilitation.

If the case for perineal closure appears clear, its execution has posed problems. Essentially, closure requires the introduction of new vascularized tissue into the defect with sufficient bulk to obliterate the dead space.

Skin grafts have a relatively limited role in groin and perineal reconstruction, being unsuitable when the bed contains infection, radiation damaged tissue, bone or prosthetic grafts. They are useful for resurfacing large degloving injuries and defects following debridement of infective lesions such as Fournier's gangrene.

There are local flaps that can be employed to resurface the perineum, including the posterior thigh flap and tensor fascia lata flap. These are not without their problems and the posterior thigh flap has an incidence of complications ranging between 20–50 per cent. The tensor fascia lata flap has a significant incidence of tip necrosis if raised to the length required to reach the perineum. Even if successful, these flaps do not address the problem of obturating the pelvic defect, preventing the bowel from settling in the pelvis.

The appropriate reconstruction, therefore, requires bulk and this can only be achieved with a myocutaneous flap. Introducing a vascularized muscle flap into the cavity has other beneficial effects. It is well known that muscle flaps bring with them a greater blood supply than cutaneous flaps. This blood supply greatly facilitates wound healing. This is probably due to transport of nutrients, oxygen, inflammatory cells and antibiotics to the local environment.[2,3] Indeed muscle flaps have been likened to 'giant macrophages' when introduced into contaminated wounds of poor vascularity.

These qualities are shared to some extent by the omentum. Omentum has been used extensively in reconstructive surgery as a pedicled flap to reconstruct the chest wall, groin and even the hand! It also has application as a free flap to provide a vascular bed in many sites around the body. It would seem logical that it has a place in the reconstruction of pelvic defects. In practice, however, it often falls short of its potential. Firstly, it may have been resected in a previous surgical procedure. Also in a debilitated patient the omentum is often thin and filmy and provides little bulk. It is also recognized that mobilizing the omentum and transposing it into the pelvis creates a 'band' that may give rise to bowel obstruction.

Until recently, most attention has been focused on the *gracilis* myocutaneous flap for the repair of perineal defects. In addition, it has been extensively used for vaginal reconstruction. Despite its widespread use, all surgeons with experience of this flap acknowledge that there are problems with reliability, particularly of the distal third. This is due to the variable position and number of pedicle vessels. Occasionally the dominant vascular pedicle is not located in the proximal portion of the muscle. Even when flap survival is total, this flap provides a disappointing volume of tissue that is usually inadequate to fulfil the obturating effect required.

The rectus abdominis myocutaneous flap

Recently, great interest has been shown in the use of the rectus abdominis flap in the reconstruction of pelvic, perineal and vaginal defects. The rectus abdominis flap was first reported by Mathes and Bostwick[4] to reconstruct defects of the abdominal wall. Since then its role has been greatly expanded and it has become a workhorse in reconstructive plastic surgery. As a pedicled flap based superiorly on the superior epigastric vessels it is used for breast and chest wall reconstruction (e.g., closure of sternotomy wounds). Based inferiorly on the inferior epigastric vessels and taking a paddle of epigastric skin, it has been used for closure of groin defects. Recently the use of this inferiorly based flap has expanded to include reconstruction of the penis, vagina, pelvis and perineum. As a free flap it also has extensive use in breast, and head and neck reconstruction.

Anatomy

The paired rectus abdominis muscles are powerful trunk flexors. They originate from the cartilages of the fifth, sixth and seventh ribs and insert into the pubic tubercle and the pubic crest. They have a dual blood supply. Inferiorly the deep inferior epigastric artery arises from the common femoral artery and traverses medially to enter the undersurface of the muscle. This is the dominant vascular pedicle and will supply the entire muscle. The superior supply is via the superior epigastric artery, a direct continuation of the internal mammary artery. Between the two vessels a 'watershed' of anastomosing vessels exists. From the anterior surface of the muscle, perforating vessels pierce the anterior rectus sheath to supply abdominal skin. These vessels are located predominantly in the peri-umbilical region. The pattern described allows a number of different patterns of flaps to be raised. Traditionally, the two main types of flap are described by the orientation of the skin paddle; i.e. the vertical rectus abdominis flap (VRAM) or the transverse rectus abdominis flap (TRAM). However, for descriptive purposes it is more appropriate to think of these flaps as either superiorly based with a hypogastric skin paddle or inferiorly based with an upper epigastric paddle.

The superiorly based flap

This flap is theoretically and practically less robust than the inferiorly based flap. Nonetheless, it has enjoyed popularity for breast and chest wall reconstruction. The skin paddle is designed below the umbilicus but is not a low paddle, because periumbilical perforators must be included. Although only skin on the ipsilateral side of the abdominal wall is considered safe, occasionally an entire transverse paddle may survive. This is illustrated in a case of chest wall reconstruction following a massive resection of a chest wall sarcoma exposing heart and lung. The chest wall was reconstructed with Marlex mesh and covered with a large rectus flap (Fig.2.1a–d).

The inferiorly based flap

Designed with an epigastric skin paddle, this flap is based on the dominant inferior epigastric vessels and is reliable and robust. Logan and Mathes drew attention to the use of this flap to reconstruct an infected radionecrotic groin wound in 1984.[5] Ideally the skin paddle is taken from a point midway between the umbilicus and the xiphisternum, and includes the area between the midline and the costal margin (Fig.2.3a). Having mapped the skin paddle, an incision is made through skin and fat, down to the aponeurosis covering the abdominal wall muscles. A long midline incision is made to expose the linea alba, through which access to the rectus muscle is achieved. The muscle is freed from its upper insertion (at the same time dividing the superior epigastric vessels) and from within the rectus sheath, taking care not to damage the epigastric vessels on its posterior surface. The inferior epigastric artery and vein enter the muscle from the lateral side, low in the sheath – these must be sought and preserved. If the graft is to be used in the pelvis and perineum, the abdominal cavity is now opened through the midline, and the myocutaneous flap is free to be swung downwards through the pelvis after appropriate dissection. In delivery of the paddle through the pelvis, care must be taken not to pull on it as this may damage its blood supply. The length of the flap is easily enough to allow the paddle to reach the perineum, where it is sutured to the edges of the skin defect.

To close the donor site, the lateral edge of the rectus sheath can be sutured to the linea alba, thus closing the muscle defect. Provided the skin paddle is not excessively large, the cutaneous donor site can be closed primarily, without split skin grafting.

This technique can be used also to close large groin defects after extensive dissection. One of us (NW) has used this technique as follows:

A 62-year-old woman was referred from the gynaecology service. She had undergone excision of a carcinoma of the vulva. At the same time a left inguinal node biopsy was performed. Histology revealed no tumour. The biopsy site failed to heal and over the subsequent 10 days the wound drained copious amounts of lymphatic fluid (>1litre/24 hours). Examination revealed a 2 × 1 cm wound with indurated edges. A formal inguinal dissection was performed including groin skin and the resultant defect closed with an inferiorly based rectus abdominis flap. The wound promptly healed (Fig.2.2a–b).

Although there are many other local flaps that can be considered for resurfacing the groin (i.e. groin flap or tensor fascia lata flap), the rectus flap is particularly appropriate for infected or irradiated wounds by virtue of the vascularized muscle introduced into the wound. An alternative myocutaneous flap is the vastus lateralis flap.[6] However, both the dissection and the donor site are greater than for the rectus flap.

Following the report by Logan and Mathes, a great interest in the various applications of the inferiorly based flap has developed. Its use has now been described for reconstruction of high pelvic defects,[7] vulval, perineal and vaginoperineal defects,[8,9] the penis[10] and extensive hemipelvectomy defects.[11] It is the use of the flap and particularly the transpelvic application that has particular relevance in the reconstruction of extensive pelvic and perineal defects. Our experience with this type of reconstruction is illustrated by the following four case reports.

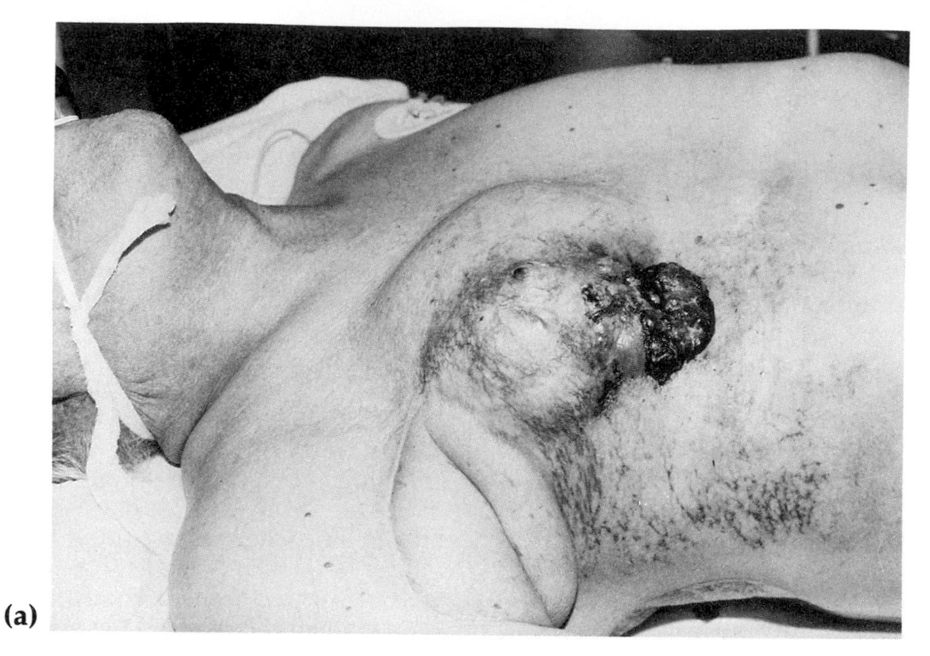

(a)

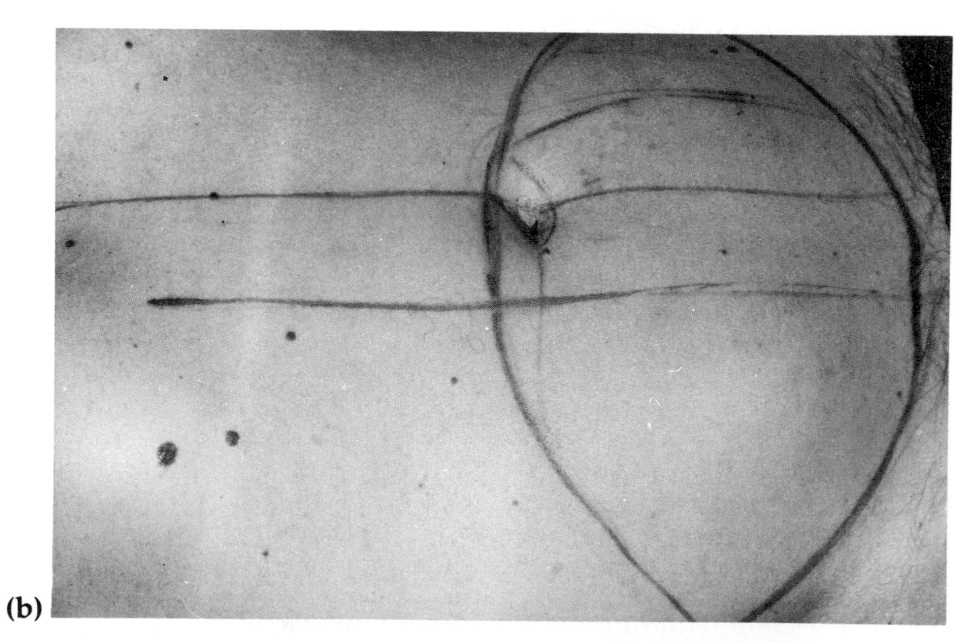

(b)

Fig. 2.1 Superiorly-based rectus abdominis flap in thoracic surgery.
(a) Soft tissue sarcoma of the right chest wall.
(b) Skin markings prior to mobilization of superiorly-based rectus abdominis flap.
(c) Chest wall defect prior to graft placement.
(d) Postoperative result.

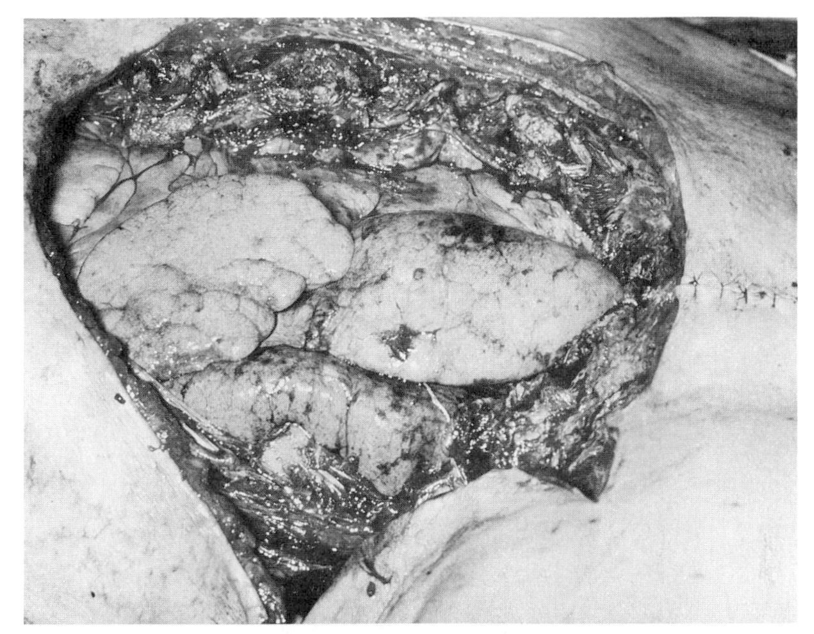

(c)

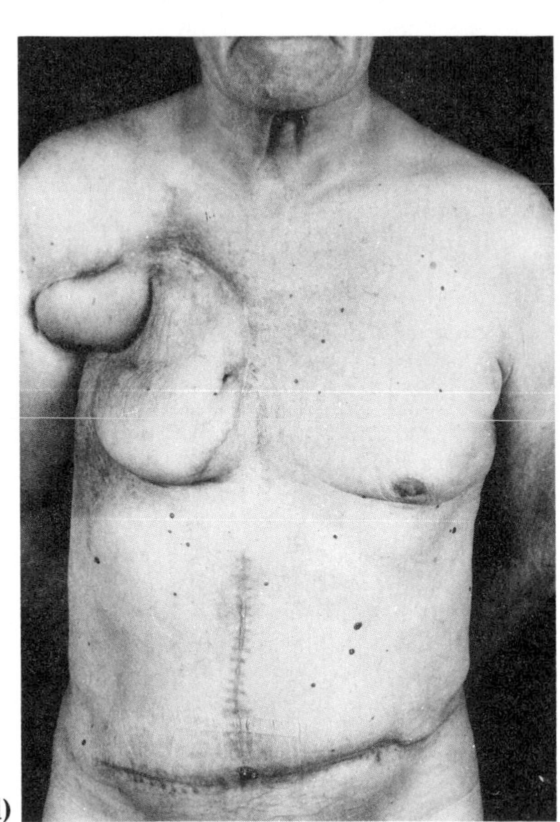

(d)

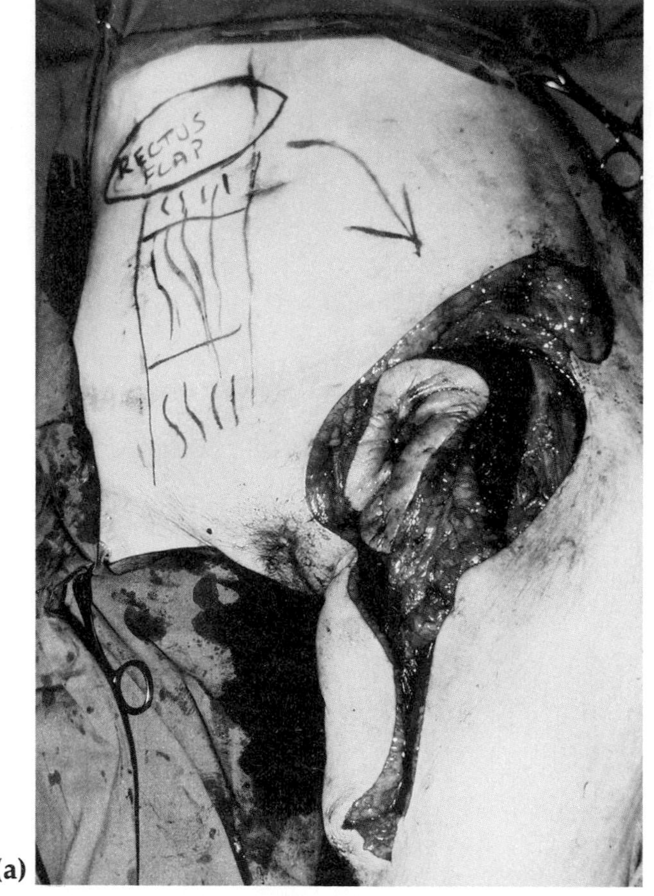

(a)

Fig. 2.2 Inferiorly-based rectus abdominis flap for use in the groin. **(a)** Malignant groin mass being mobilized prior to grafting.

CASE 1 Fig. 2.3 a–c

A 19-year-old student nurse had previously undergone a series of surgical procedures, including rectopexy and proctectomy with coloanal anastomosis, for intractable solitary rectal ulcer syndrome. Ultimately she had undergone excision of the neorectum and permanent colostomy in March 1988, but failed to heal the perineal wound. A gracilis transfer a year later failed to secure healing. In December 1989 a right rectus abdominis myocutaneous flap was tunnelled through the pelvis, with a skin paddle delivered to the perineum. Healing was prompt – the perineum remains soundly healed.

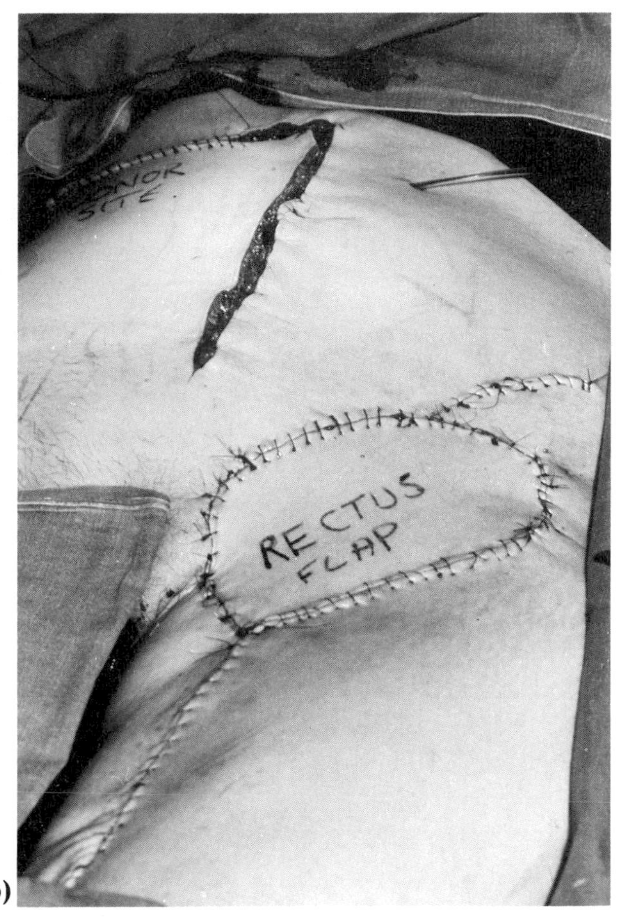

(b)

(b) After grafting with inferiorly-based flap.

CASE 2

A 26-year-old female with a long history of Crohn's disease suffered with repeated episodes of pelvic and lower abdominal sepsis related to a non-healing pelvic cavity following proctectomy. A narrow fibrous cavity extended from the perineum to the lower part of the anterior abdominal wall, and defied all conservative attempts to induce healing over a 2-year period. A right rectus abdominis flap was fashioned, which passed through the narrow, fibrotic channel with difficulty; it was not possible to deliver a skin paddle to the perineum from above, so a gracilis myocutaneous flap was used to provide skin cover for the perineum. The wounds healed with no complications, and the patient remains free of pelvic sepsis 18 months later.

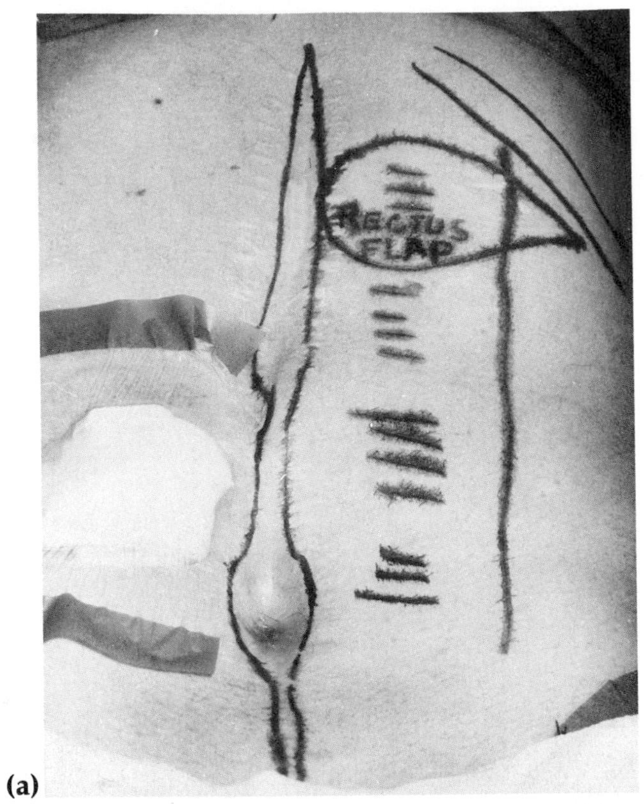

(a)

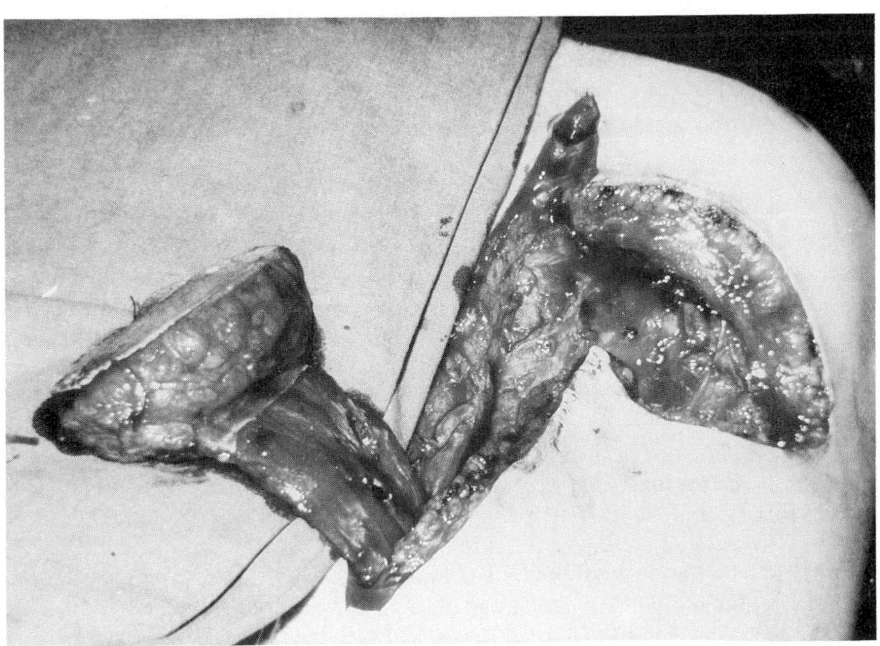

(b)

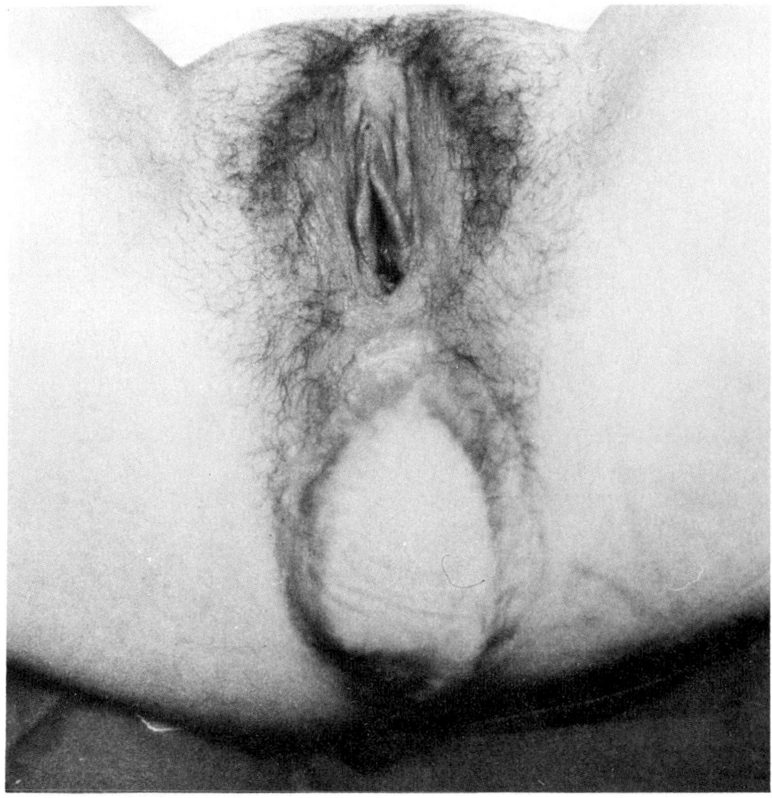

(c)

Fig. 2.3 Rectus abdominis flap in the unhealed pelvic cavity (Case 1).
(a) Skin markings prior to mobilization of inferiorly-based flap.
(b) Mobilized musculocutaneous flap prior to passage through pelvis.
(c) Postoperative appearance.

CASE 3

A 65-year-old man underwent abdominoperineal excision of the rectum with radical postoperative radiotherapy for rectal cancer. The pelvic wound failed to heal, and was lined with thick green slough typical of radionecrosis. 6 months after primary surgery a right rectus abdominis flap was tunnelled through the pelvis, with subsequent uncomplicated primary healing.

CASE 4 Fig. 2.4 a–d

A 45-year-old woman presented with a neglected squamous cell carcinoma of the vulva, extending onto the perineum, inner thighs and buttocks; the bladder neck was exposed and the rectovaginal septum destroyed. She underwent a pelvic exenteration with a terminal colostomy and ileal conduit. The pelvic outlet was filled with a right, inferiorly based transverse rectus abdominis flap, and the residual area grafted. The wound healed without complications.

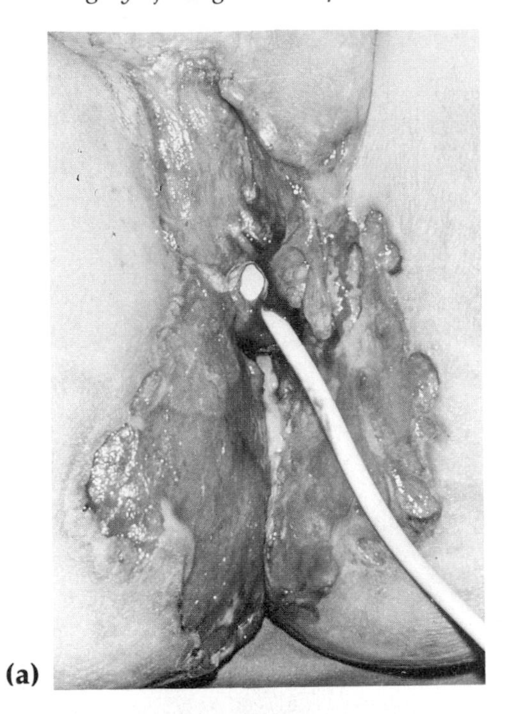

(a)

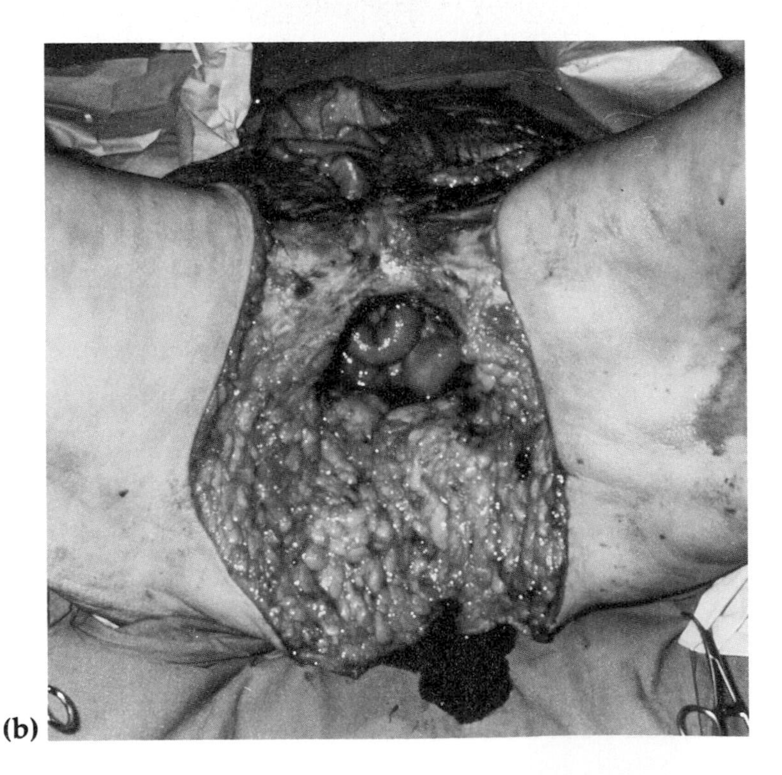

(b)

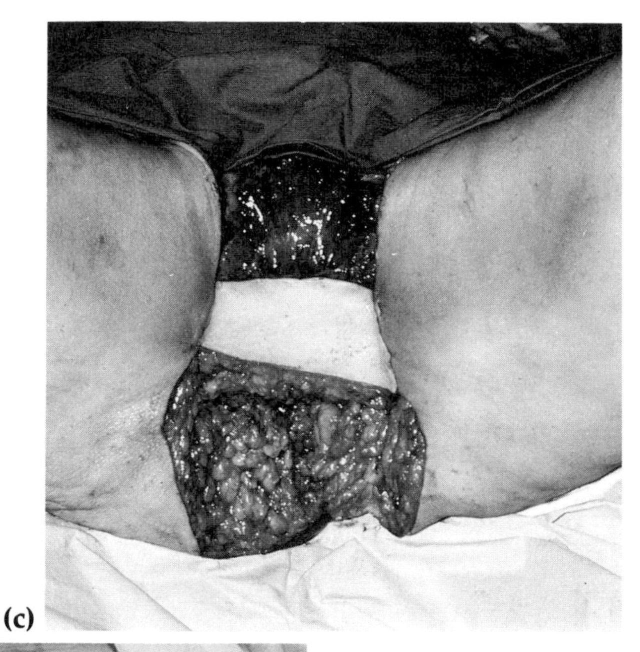

(c)

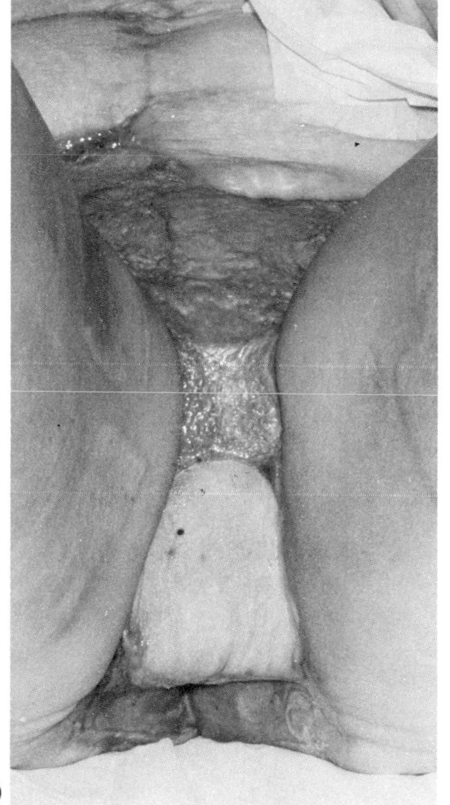

(d)

Fig. 2.4 Rectus abdominis flap in replacement of the perineum and pelvic contents.
(a) Advanced squamous cell carcinoma of the vulva.
(b) Radical clearance of pelvis and perineum.
(c) Inferiorly-based flap used to fill pelvis.
(d) Postoperative appearance. Defects not covered by flap skin have been treated using split skin grafts.

Conclusion

The rectus abdominis flap has become a workhorse in general plastic surgery. The inferiorly based flap provides a long pedicle, a large volume of muscle and a robust skin paddle. These qualities make it an invaluable flap for reconstructing the perineum pelvis and groin. For the pelvic exenteration defect it has become an automatic first choice reconstruction and has overcome many of the difficult problems of managing these defects.

Acknowledgement

The authors are grateful for the referral of case 4 to one of us (NW) by Mr N. Theodorou, Charing Cross Hospital, London.

References

1 Irvin T T, Goligher J C. A controlled clinical trial of three different methods of perineal wound management following excision of the rectum. *Br J Surg*. 1975; **62**: 287–91.
2 Jurkiewicz M, Bostwick J, Hester T, *et al*.. Infected median sternotomy wound. Successful treatment by muscle flaps. *Ann Surg*. 1980; **191**: 738–44.
3 Mathes S, Fent L, Hunt T. Coverage of the infected wound. *Ann Surg*. 1983; **198**: 420–29.
4 Mathes S J, Bostwick J, III. A rectus abdominal myocutaneous flap to reconstruct abdominal wall defects. *Br J Plast Surg*. 1977; **30**: 282–3.
5 Logan S E, Mathes S J. The use of a rectus abdominis flap to reconstruct a groin defect. *Br J Plast Surg*. 1984; **37**: 351–3.
6 Waterhouse N, Healy C. Vastus lateralis myocutaneous flap for reconstruction of defects around the groin and pelvis. *Br J Surg*. 1990; **77**: 1275–7.
7 McCraw J, Kemp G, Given F, Horton C E. Correction of high pelvic defects with the inferiorly based rectus abdominis myocutaneous flap. *Clin Plast Surg*. 1988; **15**: 449–53.
8 Skene A I, Gault D T, Woodhouse C R J, *et al*.. Perineal vulval and vaginoperineal reconstruction using the rectus abdominis myocutaneous flap. *Br J Surg*. 1990; **77**: 635–7.
9 Tobin G R, Day T G. Vaginal and pelvic reconstruction with distally based rectus abdominis myocutaneous flaps. *Plast Reconstr Surg*. 1988; **81**: 62–70.
10 Santi P, Berrino P, Canavese G, *et al*.. Immediate reconstruction of the penis using an inferiorly based rectus abdominis myocutaneous flap. *Plas Reconstr Surg*. 1988; **81**: 961–4.
11 Temple W, Naymneh W, Ketcham A. The total thigh and rectus abdominis myocutaneous flap for closure of extensive hemipelvectomy defects. *Cancer*. 1982; **50**: 2524–8.

3

Mass screening for colorectal cancer and adenomas: might it save lives?

Julia E.C.W. Verne

Introduction

This review discusses the extent to which the recommendations for introducing mass population screening are fulfilled for colorectal neoplasia and the methods currently under evaluation. In particular, current data on the efficacy of mass population screening by faecal occult blood tests and flexible sigmoidoscopy are summarized with the emphasis placed on the differences between real and perceived benefits and costs of both methods. The potential merits of focusing preventative strategy among high-risk groups are discussed and the need for a better understanding of risk factors other than familial is underlined, particularly with respect to their population prevalence and predictive value for neoplasia. Finally attention is drawn to areas in which further research is needed to verify the efficacy and increase the efficiency of secondary prevention.

The principles of mass screening

Screening is defined as the presumptive identification of disease in asymptomatic individuals by the application of simple tests or examinations.[1] This definition emphasizes two important points:

- The aim is to detect disease in individuals who are at best asymptomatic (or, as is often the case in colorectal neoplasia unaware of a problem because of the non-specific nature of the symptoms)
- Screening tests are not usually diagnostic – they merely classify individuals as being more or less likely to have the disease in question. Almost always a positive screening test needs to be followed by another investigation to verify the diagnosis

Even for common diseases, enormous numbers of individuals are screened, only to find relatively few curable cases; screening tests are invariably cheap and simple methods, suitable for mass application. Inevitably there will be some misclassification of subjects, but the ideal screening test would have:

- High sensitivity for the target condition, usually cancer but possibly also a pre-malignant condition.
- High specificity to classify correctly those without disease and a positive result should be highly predictive for the presence of the target condition[2]
- Acceptability to those being screened and clinicians because without their compliance secondary prevention will make no impact, however serious the disease or good the methods

During the first years of a screening programme two early outcomes, the stage distribution of cancers detected and the case fatality, can be determined.[2,3] These outcomes are almost universally overoptimistic because of the very nature of screening that tends to biases which:[4]

- Artificially lengthen the time of survival (lead time bias)
- Select cancers with a less aggressive course (length bias) and 'pre-malignant' lesions that do not have true malignant potential (diagnostic bias).

For these reasons site-specific mortality (i.e. death rate for the disease being screened for) is the outcome parameter of choice[2] for the evaluation of population screening programmes because it is objective and relatively easy to assess.

Morbidity is not a useful measure because it is difficult to quantify and because successful screening raises the apparent incident rate of the disease, at least in the short term. Furthermore, screening may actually increase morbidity (cancers and pre-malignant lesions that would not surface clinically within the patients life-time are found and treated) although a moderate increase may be acceptable if offset by a large reduction in mortality. Randomized controlled trials, in which mortality rates are compared between groups offered screening and those not, provide the only valid estimate of the benefits to be achieved, because they avoid many of the serious limitations of other study designs, especially the biases already mentioned.[4] They also overcome the potential bias of only screening volunteers, who may differ in their risk of colorectal cancer (higher or lower) from the general population.

The definition of the target condition, which screening aims to detect, i.e. curable cancer, pre-malignant lesion or both is crucial because the efficacy and cost effectiveness hinges on this. It is said that benefits from screening only accrue to individuals correctly classified by the screening; however, if an inappropriate stage of the disease is chosen (especially if it is an early stage with a very high population prevalence and low probability of progression) this will not necessarily hold true (see below).

Colorectal neoplasia as a clinical problem

Mass screening is only appropriate in common diseases for which an early intervention exists that can be applied to improve an otherwise poor prognosis.[1,2] At first glance colorectal cancer would appear to be an ideal target for screening. It is a major cause of death – the commonest cause of cancer-related death after lung cancer in men and breast cancer and lung cancer in women, accounting for 17053 deaths in 1989.[5] Because 5-year survival remains at only 35 per cent[6] it also represents an important cause of morbidity accounting for numerous hospital admissions for non-curative surgery, chemotherapy and terminal care. The failure to improve 5-year survival reflects the limited efficacy of treatment in patients presenting symptomatically, the majority of whom have advanced disease.[7] Yet, since the beginning of the century it has been known that early stage colorectal cancer is potentially curable.[8] The discovery and development of the guaiac test, for detecting occult blood in the faeces with which early cancers could be picked-up in asymptomatic subjects[9] made this seem a realistic possibility.

Over the past two decades, two further developments have made the prevention of colorectal cancer mortality and morbidity seem an even more possible proposition:

- The adenoma–carcinoma sequence – the hypothesis suggesting that most colorectal cancers evolve from normal mucosa through a pre-malignant lesion the adenoma[10]
- The evolution of endoscopy (first rigid and later flexible); able to detect not only early stage cancers but also adenomas as small as 1–2 mm in the distal large bowel and colonoscopy as a second wave investigation and treatment

Although the vast majority of adenomas cause no problems themselves, their significance lies in the hypothesis that the incidence of colorectal cancer could be reduced by detection and prophylactic removal of adenomas. Not only might this enable a greater reduction in mortality to be achieved by increasing the therapeutic window, but also morbidity associated with surgery for early stage cancers could be avoided by the endoscopic removal of pre-malignant adenomas. This argument remains controversial for a number of reasons:

- Despite a steady accumulation of data on the molecular biology of colorectal neoplasia to support Morson's hypothesis that the majority of colorectal cancers evolve from pre-malignant adenomas, the evidence remains largely circumstantial
- Adenomas are an extremely common finding in the population over the age of 50 (12–20 per cent)[12,13]
- Important aspects of the natural history of colorectal neoplasia need clarifying, such as the proportion of cancers arising from adenomas, the distribution of the time scale for the evolution of a carcinoma and methods to distinguish those adenomas or patients bearing adenomas most likely to develop cancer

The latter point is of utmost importance, because although it is clear that detection and treatment of adenomas could prevent a proportion of invasive cancers,[14] it is equally clear from autopsy series[12] and a small number of studies following the natural history of adenomas left *in situ*, that the vast majority of adenomas do not develop malignant change over considerable periods.[15,16]

Unfortunately, with the current state of knowledge of colorectal cancer pathogenesis, finding adenomas at endoscopy is only the beginning of a complex, expensive and sometimes risky process of further investigation, treatment and lifetime follow-up. Approximately one-third of patients with adenomas detected within the sigmoid and rectum have been shown to have synchronous proximal adenomas and/or cancer.[17] Subsequently, 30–50 per cent of patients from whom adenomas have been removed will develop new ones. At present a relative dearth of data prevents reliable prediction of:

- Patients with distal adenomas likely to have synchronous proximal adenomas
- Adenoma bearers most likely to produce new ones

Subsequent risk has been shown to vary according to the number, size and histology of the original lesions.[19] Grossman *et al.*[20] lent support to this hypothesis by suggesting that individuals with only one distal adenoma less than 1 cm in size with tubular histology do not need follow-up. However, in the absence of conclusive evidence most experts recommend that all subjects with adenomas found at screening should undergo total colonoscopy to check the proximal colon at which time all polyps found should be removed, and that subsequently all adenoma bearers are placed under life-time colonoscopic surveillance.[18] As long as there is no consensus on optimal follow-up strategy, the management of all those patients found to have colorectal adenomas at sigmoidoscopic screening will continue to present formidable problems. Any potential benefit that might be gained by such intensive surveillance, in terms of reduced premature deaths or need for abdominal surgery is likely to be achieved at considerable cost.[21,22] Hoff[21] suggested that recruiting just 100 new patients for screening sigmoidoscopy every year with 4-year follow-up for those found to have adenomas would entail a 60 per cent increase in the work load after 8 years and 90 per cent if those with multiple polyps were offered 2-yearly follow-up. Moreover, unnecessary repetitive removal of polyps, which carry little risk of undergoing malignant transformation, may incur significant risks of serious complications and even death.[22] A recent cost analysis performed using data from the medical literature in a simulation model showed that if follow-up surveillance were only 50 per cent effective in reducing mortality (i.e. a 50-year-old man's cumulative risk of dying of bowel cancer would be reduced from 2.5 to 1.25 per cent) then 1131 colonoscopies would be required to prevent one cancer death, incurring 2.3 perforations, 0.17 perforation-related deaths and physician costs of $331 000.[22] Only further progress in the understanding of the normal evolution of colorectal cancer will help clarify which preclinical stage should be targeted for screen detection – early stage cancer or adenomas.

Mass screening tests

The majority (approx. 90 per cent) of individuals developing colorectal cancer are at average risk and until now no characteristics except age have been found that could be used to identify their risk pre-morbidly. 94 per cent of new cases occur over the age of 50[5] and so this is the age at which attempts at secondary prevention generally commence. Easy access to the colorectal luminal surface and its contents has led to the development of a battery of investigations that could be used in screening. Of all the screening tests evaluated over the past decade only faecal occult blood testing and flexible sigmoidocopy have shown any promise. The American Cancer Society recommends:

- Annual rectal examination for all subjects from age 40 years old
- Annual faecal occult blood tests from age 40 years old
- 3–4 yearly sigmoidoscopy after two negative examinations a year apart from age 50 years old[23]

This rectal examination has been discarded by most other organizations interested in the evaluation of screening because of its obvious limitations, namely only 10 per cent of colorectal cancers arise within reach of an examining finger. In contrast, the United Kingdom Coordinating Commitee on Cancer Research Faecal Occult Blood Working Party,[24] the Canadian Task Force on Periodic Health examinations[25] and the National Cancer Institute[26] have deferred making recommendations regarding screening for colorectal cancer on the basis that the evidence is still inconclusive.

Faecal occult blood tests (FOBT)

These have been the most widely evaluated screening tests and among them Haemoccult achieves the best compromise between sensitivity, specificity and cost.[27] However, it only detects a non-specific marker of neoplasia, the peroxidase-like activity of haematin, a breakdown product of haemoglobin. The result is that up to 50 per cent of cancers and 75 per cent of adenomas are not detected[28] either because they bleed too little, bleed intermittently or the blood is not in the form of haematin (approx. 50 per cent of caecal and rectal cancers are missed for this reason) either having been degraded beyond haematin or not yet degraded. This results in a short therapeutic window and screening must be repeated frequently, at intervals of probably not greater than 2 years.[29] Furthermore, about half of those screened with positive results do not have neoplasia.[27,28] False positives occur either because of bleeding from non-malignant conditions or because the guaiac reaction detects other substances with peroxidase activity (from red meat, dark fish, certain vegetables) in the stool. For this reason a diet avoiding these substances is usually recommended for the duration of the test.[28] Population compliance with Haemoccult screening is still a problem in the United Kingdom[28] although less so in other countries.[30] We have recently shown that this is unlikely to be related to faecal testing but rather to public knowledge about colorectal cancer and attitudes to preventive medicine. It has been estimated that Haemoccult screening could reduce colorectal cancer mortality by about 20 per cent.[31] There is currently no evidence to confirm or refute this hypothesis. Although Haemoccult itself is inexpensive (less than £1), investigation of false positives significantly inflates the cost per case of neoplasia detected.[3] Transferring the onus of performing the FOBT screening to the patient does not necessarily improve compliance amongst physicians. The American Cancer Society[27] reported a compliance rate of 48 per cent with their recommendations for FOBT screening and Hoogewerf *et al.*[32] found that only 13 per cent of family physicians complied with an invitation to participate in a screening trial. Five large trials[31, 33–36] are currently in progress looking at the efficacy of Haemoccult screening. Of these, four are randomized controlled trials and three of these are conducted in subjects randomly selected from the population. The results of these studies are summarized below:

NOTTINGHAM[31]

This is the largest randomized controlled trial. General Practice patients were randomized either to control group or to 2-yearly Haemoccult screening. Its size and follow-up (minimum 7 years) has been designed to be 80 per cent confident of detecting a 23 per cent or greater reduction in mortality. It is probably the only trial of sufficient magnitude to do this. The higher incidence of Stage D tumours in the test goup (0.2/1000 person years) compared with the control (0.15/1000 person years) means that no mortality advantage will be seen in the near future.

DENMARK[33]

The preliminary results of this study are remarkably similar to the Nottingham study. Although early mortality data are available they are of limited value because of the short follow-up.

SWEDEN[34]

The most striking feature of this study is the very high interval cancer rate of 26 per cent (symptomatic cancers appearing between screens in the study group), which makes it less likely that this study will be able to demonstrate a mortality reduction attributable to screening.

NEW YORK[35]

The aim of this study compared the efficacy of screening by sigmoidoscopy alone with that of sigmoidoscopy combined with Haemoccult. Two groups of volunteers were selected; i) those attending for annual check-ups, ii) those presenting for the first time. Problems stemming from heterogeneity and the non-randomization of the two groups of volunteers and the lack of a Haemoccult-only group will make interpretation of possible benefits difficult.

MINNESOTA[36]

In this trial volunteers were randomized to a control group and a test group offered Haemoccult annually or every 2 years. The recruits were all volunteers and this may possibly bias the results. No mortality data have yet been published.

Flexible sigmoidoscopy (FS)

The limited access of the sigmoidoscopic examination, patient compliance, a forecast shortage of specialists and the costs of follow-up have thrown doubt on the feasibility of FS screening, despite the fact that in contrast to Haemoccult, evidence exists suggesting that the incidence of rectal cancer can be reduced by endoscopic screeening.[14] Although only the sigmoid and rectum can be examined reliably with the 60 cm FS, up to 80 per cent of all cancers and 95 per cent of adenomas can be detected if screen-detected disease is followed up by diagnostic colonoscopy.[30] This is because two-thirds of neoplasia arising beyond the reach of the 60 cm FS has synchronous, distal marker neoplasia, which is detected by FS screening. The yield for cancer and adenomas is, therefore, potentially far greater in FS screening than with Haemoccult and has led to estimates that FS screening could reduce colorectal cancer mortality by up to 85 per cent.[29] There are as yet no data in the United Kingdom on compliance with FS screening; however, in other countries the compliance rate has equalled that with FOBT screening.[15,30] Although currently there are insufficient numbers of trained endoscopists to perform FS screening nationally, both general practitioners and nurses can be trained to perform 60 cm screening sigmoidoscopy to the same level of competence as gastroenterologists.[11,37] Most of the doubts about screening sigmoidoscopy arise because of the lack of analytical studies, most information coming from uncontrolled studies with no follow-up data. Evidence about the efficacy of sigmoidoscopic screening comes from two uncontrolled studies with follow-up[14,38] and one randomized trial.[39] However, all three studies have major shortcomings. Of the two uncontrolled studies the first was conducted by Hertz *et al.*[38] in 1960. In this study 26 126 primarily asymptomatic patients underwent 40 091 rigid sigmoidoscopic examinations. Cancer was detected in 0.22 per cent and of these 81 per cent were Dukes' stage A or B. Follow-up data are available on 95 per cent of the patients, 88 per cent of whom were still alive at the end of the 5-year period. These results were excellent compared with survival in unscreened populations. However, the study was subject to the biases described previously.[4]

The major study that the American Cancer Study based its recommendations on is that reported by Gilbertsen and Nelms.[14] The authors concluded that rigid sigmoidoscopic screening could reduce the incidence of rectal cancer by 85 per cent and that those detected were only minimally invasive (all but one were at Dukes' stage A). Several authors have questioned the 85 per cent reduction in incidence.[11] Firstly cancers detected on the initial screen were excluded from the analysis – these were not cancers prevented and, therefore, would have surfaced during the follow-up period. Secondly only subjects returning repeatedly for follow-up were included. Some patients may have failed to return because of the symptomatic appearance of colorectal cancer in the interim and, this could have resulted in significant underascertainment of cases. Finally, there was no control group; thus the cohort may have differed significantly from the normal Minnesota population with respect to risk for colorectal cancer.

The results of the Kaiser–Permanente Multiphasic Evaluation Study[39] is often cited as being evidence from a randomized controlled trial for the efficacy of sigmoidoscopic screening. Its aim was to test the efficacy of Multiphasic Health Checkups (MHC), and sigmoidoscopy was only one part of this. 10 713 patients were randomized to a study group having MHCs, and a control group without MHCs. After 16 years of follow-up the cumulative mortality rate for colorectal cancer was significantly lower in the study group compared with the control group (2.3 versus 5.2 deaths/1000, $p<0.05$). Re-examination of the data[40] found that;

- The sample size would have been too small to detect a significant difference
- The 50 per cent lower cumulative incidence of colorectal cancers arising within reach of the rigid sigmoidoscope found in the study group compared with the control (4.3 versus 6.7 cases/1000) could not be accounted for by increased detection
- There was no significant difference in exposure to sigmoidocopy in study and control groups (31 per cent and 26 per cent of group members having at least one sigmoidoscopy respectively)
- The improved stage distribution for the study group tumours was almost as good for those detected by symptoms as for those screen-detected.

The case for the efficacy of sigmoidoscopic screening, therefore, remains unproven. Even if such a system were introduced in the United Kingdom, current recommendations for 3–5-yearly screening plus a similar frequency for follow-up colonoscopy in all adenoma bearers would completely overburden the health service, and might be associated with significant screen-induced morbidity as a result of repeatedly treating large numbers of adenomas with low probability of progression.[21] FS screening will only be feasible if a different strategy to those currently proposed can be developed. It has been suggested that most of the benefit in terms of reduced colorectal cancer incidence, resulting from periodic screening, is in fact accrued at the initial screen.[14] If this is so, then a single high quality 60 cm flexible sigmoidoscopic screen performed at an appropriate age with a single diagnostic colonoscopy to detect and treat proximal as well as distal neoplasia might be sufficient to achieve a cost effective reduction in mortality. We are currently investigating the feasibility of this approach.

Risk factors for colorectal neoplasia and their use in screening

Efficient application of preventive strategies depends on a good understanding of the relationship between aetiology and risk of developing colorectal cancer. Where the natural history of a disease is well understood and a risk factor(s) is known to have aetiological links with the disease, it is often possible to increase the effectiveness and efficiency of secondary prevention by focusing activities on high-risk groups.[2,41] Diagnostic and treatment strategies can be more accurately tailored to the needs of these individuals, in particular tests can be more diagnostic as fewer need to be performed. Compliance among high-risk individuals is usually higher than average risk subjects because they may have greater insight regarding their risk and the chances of benefit. Most importantly, the probability of benefit increases as the risk of occult disease increases, and the cost per case detected is inversely proportional to the risk. Secondary prevention, therefore, can often be more easily justified in high-risk groups than for the average risk general population where there is a finer dividing line between benefit and cost for each individual.

At present the best characterized risk factors that have a potential application for selecting individuals for surveillance can be classified as familial or personal, although it is likely that the underlying mechanisms of susceptibility overlap in these two groups.

Familial conditions

There are three distinct categories of familial risk. The first two familial adenomatous polyposis (FAP) and hereditary non-polyposis colorectal cancer syndromes (HNPCCSs) are characterized by autosomal transmission of a highly penetrant dominant gene associated with a very high (approx. 100 per cent) risk of developing colorectal cancer at an early age. The third category covers individuals with a small number of affected close relatives in whom a genetic pattern of inheritance cannot be definitively established.

In FAP, the natural history of the disease and the efficacy of secondary prevention have been extensively studied through the efforts of registries and specialist clinics.[42,43] Until recently, affected individuals have been identified by regular sigmoidoscopic surveillance of the offspring of affected families watching for the development of myriad adenomatous polyps throughout the bowel. The age at which sigmoidoscopic screening should start and cease is controversial, ranging from 10–15 years old to start and continuing until between 45 and 65 years old.[42,43] It is important to note that up to 35–45 per cent of new cases of FAP are 'solitary',[43] i.e they have no family history. This may be the result of new mutations, illegitimacy or poor ascertainment. However, their offspring carry the same 50 per cent risk of inheriting FAP and should, therefore, be offered the same surveillance.

Recent localization and identification of a gene on chromosome 5 that is co-inherited with the phenotype for FAP in several families means[44,45] that it may soon be possible to ascertain affected status pre-phenotypically in many families, thus sparing unaffected members sigmoidoscopic surveillance and offering the possibility of antenatal diagnosis. Once patients have developed adenomas they are offered total colectomy with ileorectal anastomosis or ileoanal pouch as, if left untreated, colorectal cancer is the inevitable outcome.[24,43] Strong evidence exists that systematic case finding and treatment of affected family members can reduce not only the incidence but also mortality from colorectal cancer.[43]

The natural history of colorectal cancer in the HNPCCSs is less well understood. Likewise, at present, there are few data on the effectiveness and efficiency of secondary prevention. However, it has been suggested that the HNPCCSs are the most common verifiable risk factor for colorectal cancer, accounting for at least 6 per cent of all new cases.[46] Most authors distinguish two separate syndromes – the site-specific colon cancer syndrome (Lynch I) in which risk for colorectal cancer alone appears to be transmitted dominantly, and the cancer family syndrome (Lynch II) in which other adenocarcinomas (breast, ovarian and endometrial) are also found.[47] Both are characterized by early age onset of colorectal cancer, relative paucity of adenomas, and multiplicity, proximal location and mucinous histology of the colorectal cancers[46,47] although none of these features alone is diagnostic. Case finding depends on detection of adenomas or, more frequently, early cancer by 3–5-yearly surveillance colonoscopy from the age of 25 years onwards.[48] The absence of a diagnostic phenotype means that identifying high-risk families rests on obtaining a history of three or more affected first-degree-related members. Consequently, many at-risk individuals are not offered appropriate surveillance. The recent localization of a gene on chromosome 18, which appears to be deleted in a few HNPCCS families,[49] has not fulfilled its initial promise as a genotypic marker of suceptibility.

Although currently no conclusive evidence exists that regular surveillance of these groups will reduce either incidence or mortality from colorectal cancer, expert opinion is united in recommending that secondary prevention should be offered to definite HNPCCS families and probably to those with two first-degree relatives, or one developing colorectal cancer before the age of 40 years.[48,50] Until these syndromes can be better characterized, the importance of taking a full family history of cancer in all new cases of colorectal cancer, especially those under 50, cannot be overemphasized.

Both FAP and the HNPCCSs are best managed in special multidisciplinary clinics affiliated to registries, so that families can be given genetic counselling, surveillance and treatment for colonic and extracolonic disease. These clinics and registries provide an invaluable source of data on the natural history of the disease and enable continued systematic evaluation of the methods, effectiveness and efficiency of secondary prevention in these groups.

Often, however, subjects present with a family of just one or two first-degree relatives with colorectal cancer. Several studies have shown that this history is associated with a two to threefold increase in risk.[50,51] For many years the aetiology of this increased risk, whether resulting from shared environmental or inherited factors, has been debated. New evidence suggests that in these individuals the risk of colorectal cancer is dominantly inherited through a gene of partial penetrance that increases suceptibility to develop adenomas.[51] No marker for this gene has been found yet, so present identification of those at risk depends on the presence or absence of a family history. Up to 7 per cent of the adult population may fall into this risk group.[51] There is currently no agreement on the correct screening strategy for these patients. Grossman and Milos.[52] found that the prevalence of neoplasia among first-degree relatives of cancer cases as assessed by colonoscopy was no greater than expected in the general population. In contrast, Rozen's group,[50] screening by flexible sigmoidoscopy and FOBT, found a threefold increase in age-adjusted risk of neoplasia compared with those with no family history. Despite their differences, both groups recommend annual FOBTs and periodic flexible sigmoidoscopy from the age of 40 in the first-degree relatives of cancer patients. Of the two screening tests flexible sigmoidoscopy would appear to be the more appropriate in the light of the work of Cannon-Albright *et al.*[51] suggesting that the adenoma is the early phenotypic expression of the trait. However, because at least 10 per cent of the population over the age of 50 have at least one first-degree relative with bowel cancer, the large numbers of individuals potentially involved necessitates urgent research to characterize the distribution of risk within this group.

Personal risk factors

Symptoms

Of all the symptoms evaluated, only the observation of frank blood in the stool has any significant predictive value for the presence of neoplasia.[53–56] Moreover, earlier diagnosis of symptoms does not improve the Dukes' stage distribution of cancers[55] nor does the duration of symptoms at diagnosis correlate with survival.[57]

History of colorectal cancer or adenomas

Although this group is normally included in the high-risk category because of the documented risk of metachronous cancer,[58] in fact these are not risk factors with respect to selection for screening, but rather for follow-up surveillance.

Ulcerative colitis

The risk of developing colorectal cancer and the need for its secondary prevention has been extensively studied in ulcerative colitis sufferers, and they constitute the other high-risk group for whom secondary prevention can be currently recommended. The risk of colorectal cancer is significant only in long-standing extensive colitis;[59] thus, most experts recommend regular colonoscopic surveillance with multiple biopsies examined for dysplasia in patients with disease of greater than 8 years duration.[59]

Personal history of cancer

A personal history of breast, endometrial or ovarian cancer[60–62] has been shown to increase moderately (two to threefold) the future risk of developing colorectal cancer. In patients having the first cancer treated under the age of 50 the risk of future development of colorectal cancer may be significantly higher (increased up to eight times).[61] Although these cancers are fairly prevalent in the general population the risk conveyed is difficult to quantify and the aetiological basis for it is poorly understood. Thus, although a few centres offer special surveillance to these groups[60] there is insufficient evidence to make firm recommendations.

Other risk factors

The increased risk associated with many other conditions, for example cholecystectomy, gastrectomy, nulliparity and atherosclerosis have been debated but none of these have been shown conclusively to convey an increased risk. It has been suggested that the cumulative risk from a combination of predisposing factors may have greater ability to discriminate those at risk. Hoff and Larsen[63] attempted to do this by collating a large amount of clinical and case history data. Using discriminant analysis they were able to classify correctly 75 per cent of individuals into a group with adenomas greater than 5 mm or to a group without. Some authors believe that focusing prevention on high-risk groups will not be feasible on the grounds that the majority of patients do not come from a specific high-risk group.[55] Until more data become available on the prevalence of risk factors in the population and their predictive value for neoplasia, this decision should be deferred.

To summarize, at present just under 10 per cent of new cases of colorectal cancer develop in very high-risk groups. Asymptomatic members of these groups should be offered special surveillance. There is insufficient data to make firm recommendations about other groups whose risk, though moderately increased, is not yet known to have a causal relationship, and which can not be quantified accurately.

Ethical issues

It will be apparent from the discussion so far that mass screening involves the participation of very large numbers of subjects and their clinicians at considerable expense and often inconvenience. For screening to be justifiable, a disease must impose a significant clinical burden on the community. Mass screening for colorectal neoplasia aims to improve the health of the community by trying to achieve an optimal balance between benefits (reducing mortality and morbidity) and costs. It will certainly not benefit every individual within that community. Indeed, screening may actually increase morbidity (at least initially) by uncovering occult disease that would either not cause morbidity or mortality or for which the prognosis cannot be improved.[2] Health service providers, therefore, have an ethical responsibility to ensure that the overall benefits to the community outweigh the costs, both financial and intangible. To assist in their decision making, certain criteria should be fulfilled:

- The disease poses an important health problem
- The natural history of the disease is well understood
- Prognosis can be improved in a significant proportion of individuals
- The risks of physical or psychological harm from the programme are less than the likely benefits
- There are sufficient resources to meet the increased demand that may be generated by screening without diverting them from perhaps more important aspects of health care
- The methods are acceptable to patients and clinicians

Furthermore, if such evidence is currently unavailable, screening should only be offered in the context of properly designed trials.

Conclusions

There can be little debate that colorectal cancer poses an important health problem but most of the other criteria for recommending mass population screening remain unfulfilled. Several randomized trials of the efficacy of FOBT are in progress and their results are awaited. Analytical studies on the efficacy of sigmoidoscopic screening are still urgently needed. Several recommendations can be made about the design of future trials:

- They should be conducted in the setting and by the personnel ultimately most likely to perform screening. Trials conducted by specialists in specialized centres are not a good model for the practical introduction of preventive measures
- They should incorporate supplementary studies on the prevalence and predictive value for neoplasia of putative risk factors. This information will enable future efforts to be restricted to individuals most at risk, thereby avoiding the discomfort, anxiety and unnecessary risk of screening in low-risk individuals
- More data should be collected on the intangible benefits and costs of screening to permit more comprehensive cost analyses to be performed for proposed screening strategies

Two other areas deserve special attention, that of the problem of risk attribution in patients with a family history inadequate to diagnose HNPCSs and in patients with a previous history of adenomas. Undoubtably risk in both groups is heterogenous. An improved understanding of its distribution in these groups will permit more appropriate allocation of surveillance methods and resources.

References

1 Wilson J M J, Jungner G. *Principles and practice of screening for cancer*. Geneva: WHO, 1968.
2 Cole P, Morrison A S. Basic issues in population screening for cancer. *J Natl Cancer Inst*. 1980; **64**: 1263–72.

3 Simon J B. Occult blood screening for colorectal carcinoma: a critical review. *Gastroenterology*. 1985; **88**: 820–37.
4 Chuong J J H. A screening primer: basic principles, criteria, and pitfalls of screening with comments on colorectal carcinoma. *J Clin Gastroenterol*. 1983; **5**: 229–33.
5 Office of Population Censuses and Surveys. *Mortality Statistics, Cancer* (Series DH2, no. **14**). London: HMSO, 1989.
6 Office of Population Censuses and Surveys. *Cancer Survival, 1979–81 Registrations* (Series MB1, **86/2**). London: HMSO, 1986.
7 Stower M J, Hardcastle J D. Five year survival of 115 patients with colorectal cancer over an 8-year period in a single hospital. *Eur J Clin Oncol*. 1985; **11**: 119–23.
8 Cloggs H S. Cancer of the colon. A study of 72 cases. *Lancet*. 1908; **ii**: 1007–12.
9 Greegor D H. Occult blood testing for detection of asymptomatic colon cancer. *Cancer*. 1971; **28**: 131–4.
10 Morson B C. The polyp cancer sequence in the large bowel. *Proc R Soc Med*. 1974; **34**: 845.
11 Selby J V, Friedman G D. Sigmoidoscopy in the periodic health examination of asymptomatic adults. *JAMA*. 1989; **261**: 594–601.
12 Williams A R, Balasooriya B A W, Day D W. Polyps and cancer of the large bowel: a necroscopy study in Liverpool. *Gut*. 1982; **23**: 835–42.
13 Hoff G, Vatn M, Gjone E, *et al*.. Epidemiology of polyps in the rectum and sigmoid colon; design of a population screening study. *Scand J Gastroenterol*. 1985; **20**: 351–5
14 Gilbertsen V A, Nelms J M. The prevention of invasive cancer of the rectum. *Cancer*. 1978; **41**: 1137–9.
15 Hoff G, Foerster A, Vatn M H, *et al*.. Epidemiology of polyp in the rectum and colon. Recovery and evaluation of unresected polyp 2 years after detection. *Scand J Gastroenterol*. 1986; **21**: 853–62.
16 Stryker S, Wolff B G, Culp C E, *et al*.. Natural history of untreated colonic polyps. *Gastroenterology*. 1987; **93**: 1009–1013.
17 Tedesco F J, Waye J D, Avella J R, Villalobos M M. Diagnostic implications of the spatial distribution of colonic mass lesions (polyps and cancers). A prospective colonoscopic study. *Gastrointest Endosc*. 1980; **26**: 395–7.
18 Winawer S J, Ritchies M T, Diaz B J *et al*.. The National Polyp Study: aims and organization. *Front Gastrointest Res*. 1986; **10**: 216–25.
19 Morson B C, Bussey H J R. Magnitude of risk for cancer in patients with colorectal adenomas. *Br J Surg*. 1985; **72**: S23–5.
20 Grossman S, Milos M, Tekawa I S, *et al*.. Colonoscopic screening of persons with suspected risk factors for colon cancer: II. Past history of colorectal neoplasms. *Gastroenterology*. 1989; **96**: 299–306.
21 Hoff G. Colorectal polyps. Clinical implications: screening and cancer prevention. *Scand J Gastroenterol*. 1987; **22**: 769–75.
22 Ransohoff D F, Lang C A, Kuo H S. Colonoscopic surveillance after polypectomy: considerations of cost effectiveness. *Ann Intern Med*. 1991; **114**: 177–82.
23 American Cancer Society. Guidelines for the cancer-related check-up. *Cancer*. 1985; **30**: 194–240.
24 UKCCCR Report of United Kingdom Coordinating Committee on Cancer Research Working Party on Faecal Occult Blood Testing. Medical Research Council: London, 1989.
25 Canadian task force on periodic health examination. *Can Med Assoc J*. 1979; **121**: 1–46.

26 Cancer control objectives for the nation: 1985–2000. *Natl Cancer Inst Monogr*. 1986; 27–42.
27 Knight K K, Fielding J E, Battista R. Occult blood screening for colorectal cancer. *JAMA*. 1989; **261**: 587–93.
28 Hardcastle J D, Pye G. Screening for colorectal cancer: a critical review. *World J Surg*. 1989; **13**: 38–44
29 Eddy D M, Nugent W, Eddy J F, *et al.*. Screening for colorectal cancer in a high risk population. *Gastroenterology*. 1987; **92**: 682–92.
30 Rozen P, Ron E, Fireman, *et al.*. The relative value of faecal occult blood tests and flexible sigmoidoscopy in screening for large bowel neoplasia. *Cancer*. 1987; **60**: 2553–8.
31 Hardcastle J D, Thomas W M, Chamberlain J, *et al.*. Randomized, controlled trial of faecal occult blood screening for colorectal cancer. Results for first 107/349 subjects. *Lancet*. 1989; **i**: 1160–4.
32 Hoogewerf P, Hislop T G, Morrison B J, *et al.*. Patient compliance with screening for fecal occult blood in family practice. *CMAJ*. 1987; **137**: 195–8.
33 Krongberg O, Fenger C, Sondergaard O, *et al.*. Initial mass screening for colorectal cancer with fecal occult blood test. A prospective randomized study at Funen in Denmark. *Scand J Gastroenterol*. 1987; **22**: 677.
34 Kewenter J, Bjork S, Haglind E, *et al.*. Screening and rescreening for colorectal cancer: a controlled trial of faecal occult blood testing in 27 700 subjects. *Cancer*. 1988; **62**: 645–51.
35 Winawer S J. Detection and diagnosis of colorectal cancer. *Cancer*. 1983; **51**: 2519–24.
36 Nivatvongs S, Gilbertsen V A, Goldberg S M, *et al.*. Distribution of large bowel cancers detected by occult blood test in asymptomatic patients. *Dis Col Rectum*. 1982; **25**: 420.
37 Rosevelt J, Frankl H. Colorectal cancer screening by nurse practitioners using 60 cm flexible sigmoidoscope. *Dig Dis Sci*. 1984; **29**: 161–3.
38 Hertz R E L, Deddish M R, Day E. Value of periodic examinations in detecting cancer of the colon and rectum. *Postgrad Med*. 1960; **27**: 290–4.
39 Friedman G D, Collen M F, Fireman B H. Multiphasic health check-up evaluation: a 16-year follow-up. *J Chronic Dis*. 1986; **39**: 453–63.
40 Selby J V, Friedman G D, Collen M F. Sigmoidoscopy and mortality from colorectal cancer: The Kaiser Permanente Multiphasic Evaluation Study. *J Clin Epidemiol*. 1988; **41**(5): 427–34.
41 Morrison A S. The public health value of using epidemiology information to identify high-risk groups for bladder cancer screening. *Semin Oncol*. 1979; **6**: 184–8.
42 Bussey H J R. Familial polyposis coli. *Pathol Annu*. 1979; **14**: 61–81.
43 Jarvinen H J, Husa A, Aukee S, *et al.*. Finnish registry for familial adenomatosis coli. *Scand J Gastroenterol*. 1984; **19**: 941–6.
44 Kinzler K W, Nilbert M C, Su L-K, *et al.*. Identification of FAP locus genes from chromosome 5q21. *Science*. 1991; **253**: 661–5.
45 Groden J, Thliveris A, Samowitz W, *et al.*. Identification and characterization of the familial adenomatous polyposis coli gene. *Cell*. 1991; **66**: 589–600.
46 Mecklin J-P. Frequency of hereditary colorectal carcinoma. *Gastroenterology*. 1987; **93**: 1021–5.
47 Lynch H T, Kimberling W, Albano W A, *et al.*. Hereditary nonpolyposis colorectal cancer (Lynch syndrome I and II). *Cancer*. 1985; **56**: 934–8.
48 Houlston R S, Murday V, Haracopos C, *et al.*. Screening and genetic councelling for relatives of patients with colorectal cancer in a family cancer clinic. *Br Med J*. 1990; **301**: 366–8.

49 Fearon E R, Cho K R, Nigro J M, *et al.*. Identification of a chromosome 18q gene that is altered in colorectal cancers. *Science*. 1990; **247**: 49.
50 Rozen P, Fireman Z, Figer A, *et al.*. Family history of colorectal cancer as a marker of potential malignancy within a screening program. *Cancer*. 1987; **60**: 248–54.
51 Cannon-Albright L A, Skolnick M H, Bishop T, *et al.*. Common inheritance of susceptibility to colonic adenomatous polyp and associated colorectal cancers. *N Engl J Med*. 1988; **319**: 533–7.
52 Grossman S, Milos M. Colonoscopic screening of people with suspected risk factors for colon cancer: I. Family history. *Gastroenterology*. 1988; **94**: 395–400.
53 Chapuis P H, Goulston K J, Dent O F, Tait A D. Predictive value of rectal bleeding for rectal and sigmoid polyps. *Br Med J*. 1985; **290**: 1546–8.
54 Silman A J, Mitchell P, Nicholls R J, *et al.*. Self-reported dark red bleeding as a marker comparable with occult blood testing in screening for large bowel neoplasms. *Br J Surg*. 1983; **70**: 721–4.
55 Farrands P A, Hardcastle J D. Colorectal screening by self-administered questionnaire. *Gut*. 1984; **25**: 445.
56 Castiglione G, Ciatto S. Selection criteria in colorectal cancer screening. *Tumori*. 1988; **74**: 451–6.
57 Holliday H W, Hardcastle J D. Delay in the diagnosis and treatment of symptomatic colorectal cancer. *Lancet*. 1979; **i**: 309.
58 Winawer S J, Sherlock P. Surveillance for colorectal cancer in average risk patients, familial high-risk groups, and patients with adenomas. *Cancer*. 1982; **50**: 2609–614.
59 Lennard-Jones J E, Morson B C, Ritchie J K, Williams C B. Cancer surveillance in ulcerative colitis. Experience over 15 years. *Lancet*. 1983; **16**: 149–52.
60 Rozen P, Fireman Z, Figer A, *et al.*. Colorectal tumour screening in women with a past history of breast, uterine, or ovarian malignancies. *Cancer*. 1986; **57**: 1235–9.
61 Bremond A, Collet P, Lambert R, Martin J L. Breast cancer and polyps of the colon. A case control study. *Cancer*. 1984; **1**: 2568–70.
62 Giacosa A, Sukkar S G, Frascio F. Surveillance of high risk patients for colorectal cancer. In Fairvre J, Hill M J (Eds), *Causation and prevention of colorectal cancer*. Amsterdam: Elsevier Science Publishers, 1987, pp. 195–204.
63 Hoff G, Larsen S. Epidemiology of polyps in the rectum and sigmoid colon. Discriminant analysis for identification of individuals at risk of developing colorectal neoplasia. *Scand J Gastroenterol*. 1986; **21**: 848–52.

4

Surgery for fistula-in-ano. Can we conserve the sphincters?

Peter J. Lunniss and Robin K.S. Phillips

Introduction

Epidemiological studies of non-specific fistula-in-ano indicate an incidence in Northern Europe of approximately 10:100 000.[1,2] No relation between complexity of fistula and racial origin has been found.[3]

Fistulae are more common in men than in women, and in infants only baby boys are affected. Attempts to preserve sphincter function would be helped if it were understood what makes the male sex prone to fistula formation; but short of such a fundamental understanding in aetiology, attention needs to be directed towards treatment of established anal fistulae in a more 'conservative' way.

Possible approaches include the use of appropriate antibiotics, and various strategies for preserving external, and indeed, internal anal sphincter function.

Little is known about the microbiological flora of non-specific fistulae. Certainly *Mycobacterium tuberculosis* can cause fistulae, and treatment of tuberculosis with chemotherapy leads to healing of the fistula. Organisms in non-specific fistula are sparse and of faecal type. It is likely that many acute pyogenic infections are treated successfully by drainage and/or antibiotics without subsequent problems with a fistula. Indeed, Grace *et al.*[4] found that in only 59 per cent of abscesses that grew gut-related organisms, could a fistula be demonstrated at the time of presentation or subsequently; presumably the other 41 per cent had a communication that healed spontaneously.

If antibiotics are to be used to treat fistula-in-ano, it will be necessary to monitor fistula healing accurately, including detection of pockets or extensions of pus that might cause recrudescence. At present, the choice lies between clinical examination, ultrasound, fistulography, computed tomography (CT) scanning and magnetic resonance imaging (MRI). Of these, we have found MRI to be the most promising (Fig. 4.1). Its advantage over ultrasound lies in its ability to detect sepsis outside the sphincters (Table 4.1).

When can the internal anal sphincter be preserved in fistula surgery? A

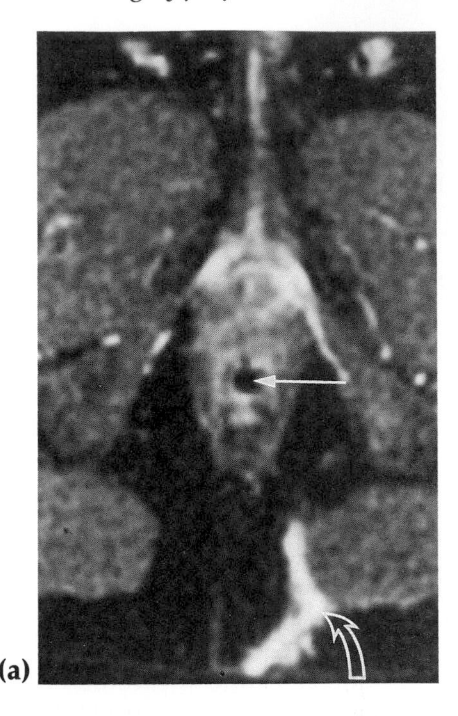

(a)

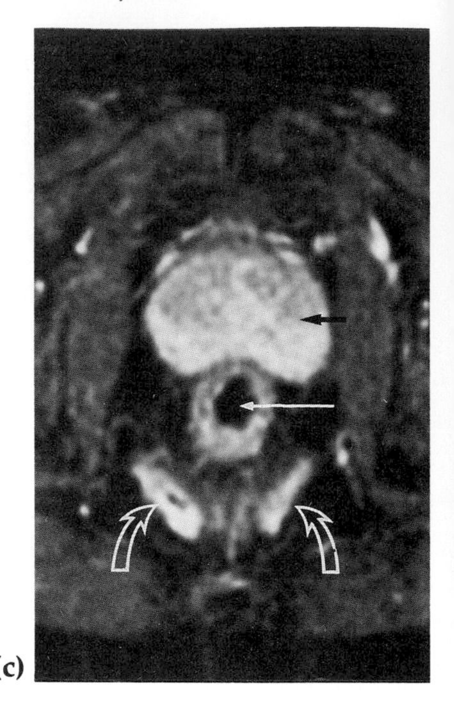

(c)

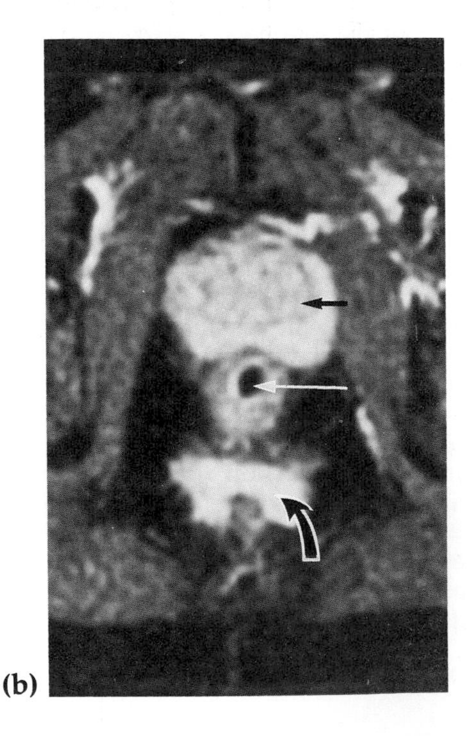

(b)

Fig. 4.1 Photographs of magnetic resonance imaging scans (STIR sequences, axial sections) of a complex fistula-in-ano. Granulation tissue and pus (open white arrows) are seen brightly on a dark background of fat. The upper photograph **(a)** shows part of the fistula track passing through the ischiorectal fat to the external opening. A supralevator horseshoe extension is seen in the middle photograph **(b)**, which extends upwards and forwards on each side of the pararectal space **(c)**. The prostate gland (dark arrows) is seen anteriorly, lying in front of the anorectal lumen (white arrows).

Table 4.1 Comparison between magnetic resonance imaging (MRI) and operative findings in the first ten consecutive cases of fistula-in-ano assessed by MRI

	MRI	Operation
Primary track		
Superficial	1	1
Intersphincteric	0	0
Transsphincteric	6	6
Suprasphincteric	0	0
Extrasphincteric	3	3
Extensions/abscess		
Infralevator	3	3
Supralevator	5*	4
Horseshoeing		
Intersphincteric	1	1
Infralevator	0	0
Supralevator	2	2

*One patient had induration above the levators palpable at surgery (as had been demonstrated by MRI) but this was not considered to be significant enough to warrant intervention.

key factor may be whether intersphincteric sepsis is always present, thereby requiring routine 'laying open' of the intersphincteric space by internal sphincter division; if intersphincteric sepsis is not always present, it may be that the internal sphincter could be preserved in some cases. Both Parks and Eisenhammer argued that cryptoglandular infection is central to the development of fistula-in-ano; in fact their classifications are based on this premise. By their reasoning one of the anal glands that lies at the level of the anal valves becomes infected; then sepsis may track in various directions to present acutely as a perianal or an ischiorectal abscess, or more insidiously as a fistula. In 1967, Goligher *et al.*[5] dissected the intersphincteric space in 28 patients presenting acutely with perineal sepsis and found an intersphincteric abscess in only 29 per cent; moreover, in only 44 per cent cases of chronic fistula did evidence of intersphincteric space involvement exist. This is the only study so far that casts doubt on the cryptoglandular hypothesis; for the most part surgeons are happy to accept the hypothesis and the classification and treatment strategies that follow from it.

Classification

The relative incidence of different types of fistula is dependent upon the classification used. That employed most widely in the United Kingdom was proposed by Parks, Gordon and Hardcastle[6] who studied 400 cases treated at St Mark's. It assumes that the primary site of infection is within the intersphincteric space and divides fistulae into four main groups: intersphincteric; transsphincteric; suprasphincteric and; extrasphincteric, according to the course of the primary track in relation to the anorectal sphincter complex. Sepsis found above the levator ani muscles must arise from any one of three sources:

- Pelvic disease
- Extension upwards of an intersphincteric fistula
- Extension upwards of a transsphincteric fistula

Parks stressed the importance of accurate determination of the origin of the sepsis before surgical intervention because:

- Drainage of a pelvic abscess through the ischio-rectal fossa would create an extrasphincteric fistula, which if laid open in the traditional manner would result in total incontinence (Fig. 4.2).
- Drainage of a supralevator extension of an intersphincteric abscess through the ischio-rectal space rather than into the rectum would result in a suprasphincteric fistula (Fig. 4.3).
- Drainage of a supralevator extension of a transsphincteric fistula must be via the perineum as drainage into the rectum could result in an extrasphincteric fistula (Fig. 4.4).

Although the St Mark's classification is based mainly on vertical and lateral extension, circumferential extension is also noted that may be in any of the three planes: intersphincteric, ischio-rectal or para-rectal (Fig. 4.5). Marks and Ritchie[3] used the same classification in their analysis of

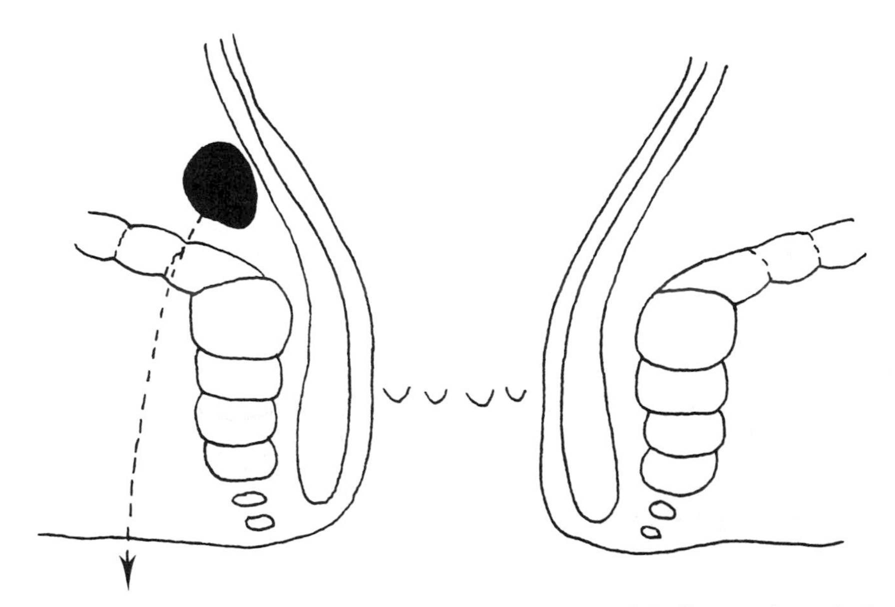

Fig. 4.2 Drainage of a pelvic abscess originating in pelvic disease through the perineum may well cause an extrasphincteric fistula.

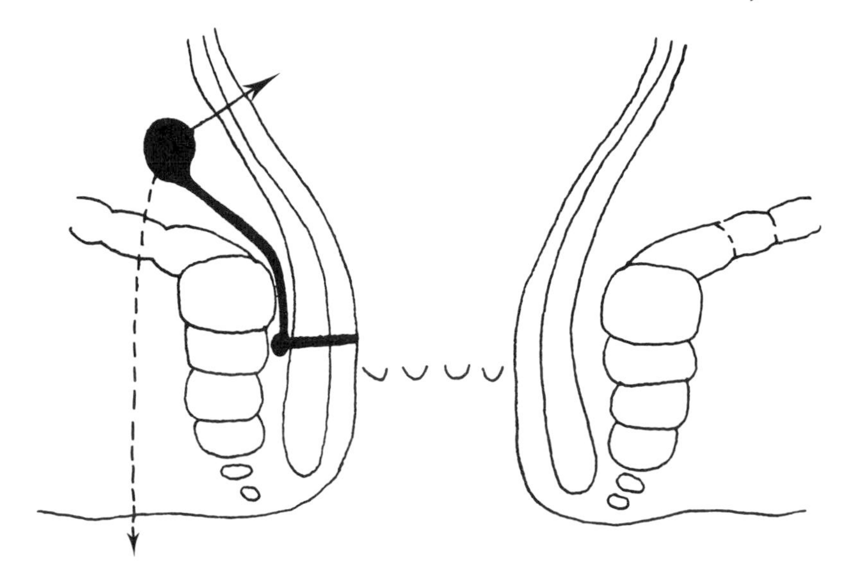

Fig. 4.3 A pararectal abscess due to a high intersphincteric extension can be drained safely into the rectum but not through the ischiorectal fossa.

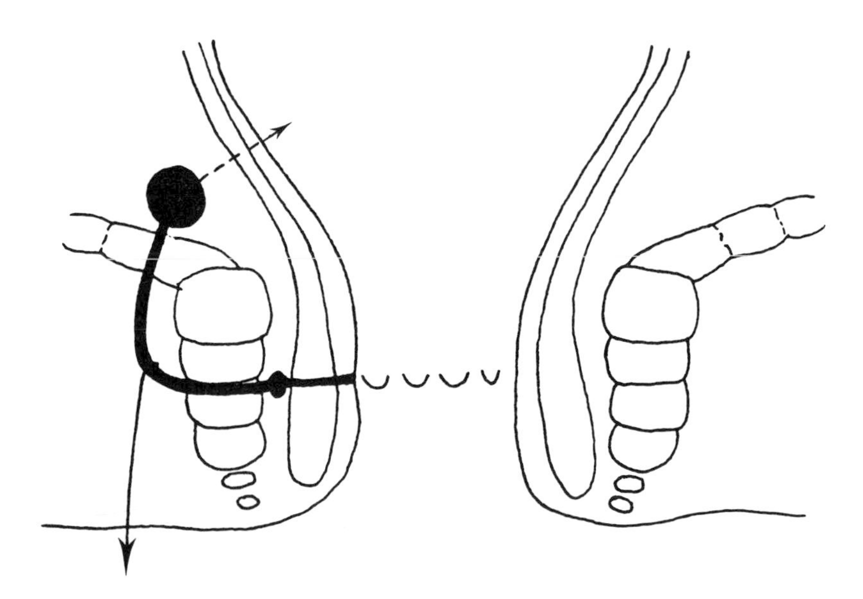

Fig. 4.4 A pararectal abscess due to extension of a transsphincteric fistula through the levator plate must be drained through the perineum and not into the rectum.

Table 4.2 The St Mark's classification applied to 793 consecutive patients presenting to St Mark's (1968–73) with idiopathic fistula-in-ano

Type of fistula	Number of patients			%
Superficial		126	520	65.6
Subcutaneous	54			
With fissure	56			
Surgical wound	16			
Intersphincteric				
Simple		394		
With extension below puborectalis		22	36	4.5
With extension above puborectalis		11		
With entry into rectum		3		
Transsphincteric				
Simple		56	164	20.7
With ischiorectal fossa extension		87		
With supralevator extension		21		
Suprasphincteric				
Simple		5	26	3.3
With ischiorectal fossa extension		1		
With supralevator extension		20		
Extrasphincteric		23		2.9
Multiple		17		2.1
Unclassified		7		0.9

Reproduced by permission of the publishers, Butterworth-Heinemann Ltd, from Marks and Ritchie.[3]

793 consecutive patients at St Mark's and reported that 66 per cent were superficial or simple intersphincteric tracks, 21 per cent transsphincteric, 3.3 per cent suprasphincteric and 2.9 per cent extrasphincteric (Table 4.2). This differs from published data from other centres, in which there is a much higher proportion of simple fistulae.[7,8]

Eisenhammer[9] acknowledged the St Mark's classification but felt that the pathology of the primary cryptoglandular fistulous abscess groups followed a consistent and predetermined anatomical path and that the relatively high incidence of unusual and complex fistulae seen at St Mark's was because of surgical intervention. In his classification, the low intermuscular abscess fistula represented 85 per cent of the cryptoglandular anorectal suppurative conditions (Fig. 4.6). The intermuscular transsphincteric ischiorectal fistulous abscess represented 13 per cent (Fig. 4.7), and the high intermuscular fistulous abscess 2 per cent (Fig. 4.8). He placed all those conditions that he felt were not of cryptoglandular origin into the heterogenous group called the acute anorectal non-cryptoglandular non-fistulous abscesses (Fig. 4.9). Thus, although Eisenhammer[10] agreed with Parks[11] concerning the origin of the fistula, he considered complex fistulae to be the result of faulty initial surgery or persistence in conservative antibiotic therapy that allowed spread of sepsis. He also noted that most anorectal fistula work in England is performed by general surgeons, St Mark's being

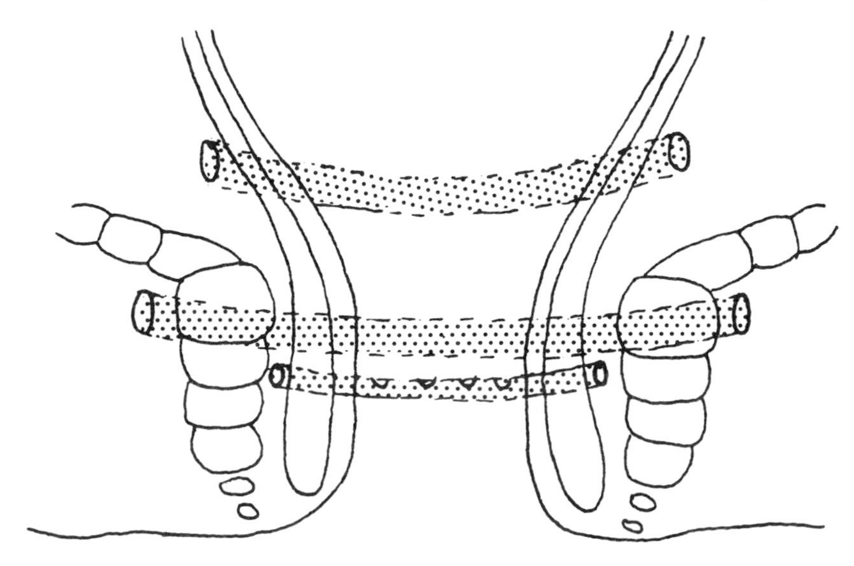

Fig. 4.5 Diagram to show the three planes in which circumferential spread can occur.

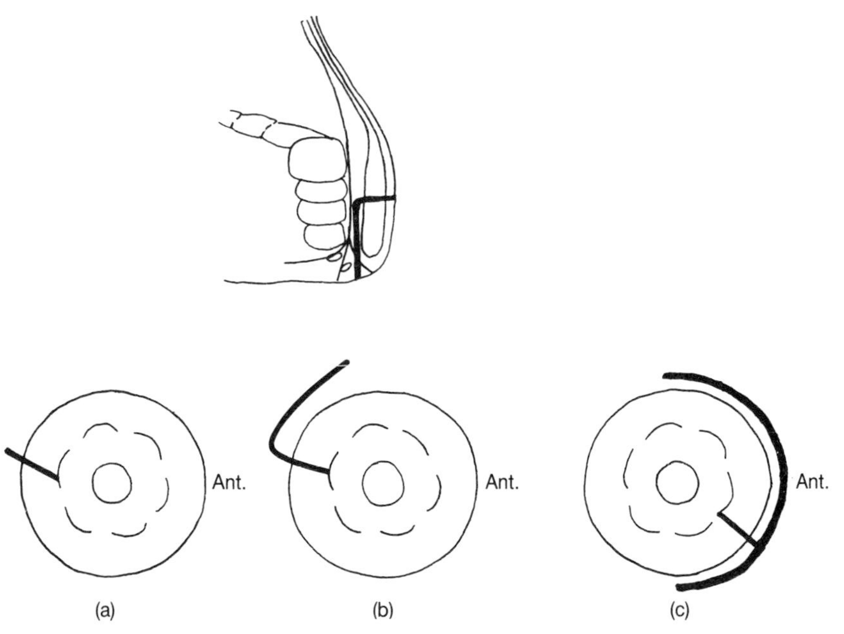

Fig. 4.6 The low intermuscular fistulous abscess and fistula of Eisenhammer. Adapted from Eisenhammer S.[9] 90 per cent occur posteriorly **(a)**, and may rarely spread anteriorly in the superficial ischiorectal space as a unilateral horseshoe **(b)**. The remaining 10 per cent occur anteriorly, and very rarely may spread bilaterally **(c)**.

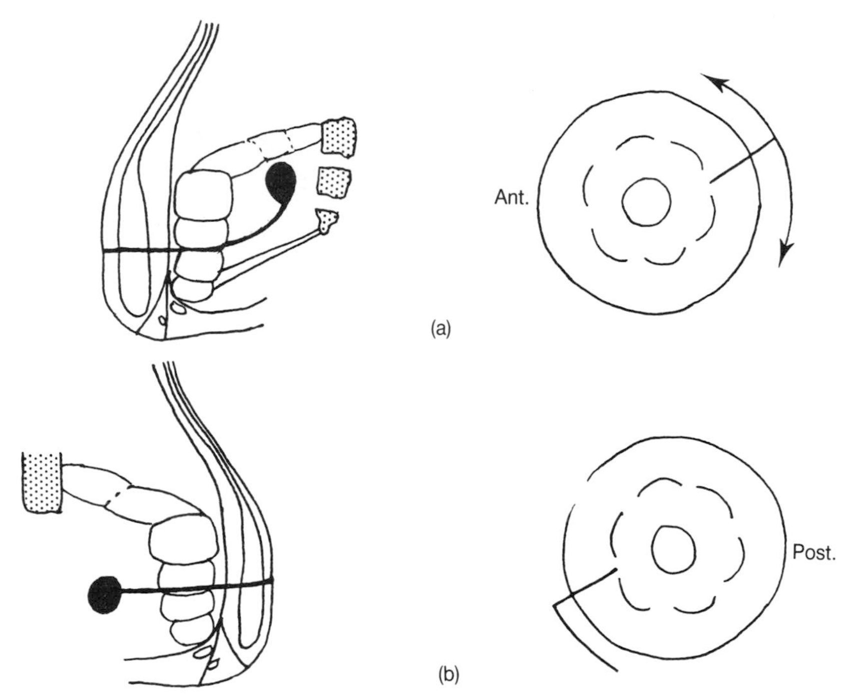

Fig. 4.7 The intermuscular transsphincteric ischiorectal fistula of Eisenhammer. Adapted from Eisenhammer S.[9] 90 per cent occur posteriorly, the abscess developing in the deep post-anal space and spreading initially unilaterally but later bilaterally **(a)**. The corresponding anterior lesion is more superficial and only spreads in one direction **(b)**.

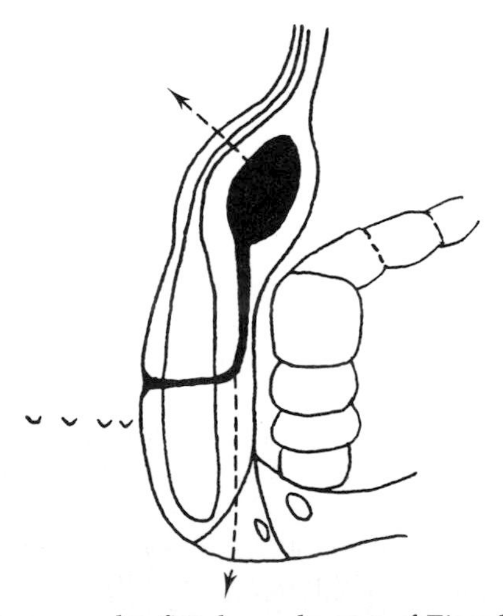

Fig. 4.8 The high intermuscular fistulous abscess of Eisenhammer. Adapted from Eisenhammer S.[9] This posterior lesion arises from an anal gland situated above the level of the dentate line, and spreads upwards in the intermuscular space above the anorectal ring. Spontaneous rupture (dotted lines) will form the corresponding fistula.

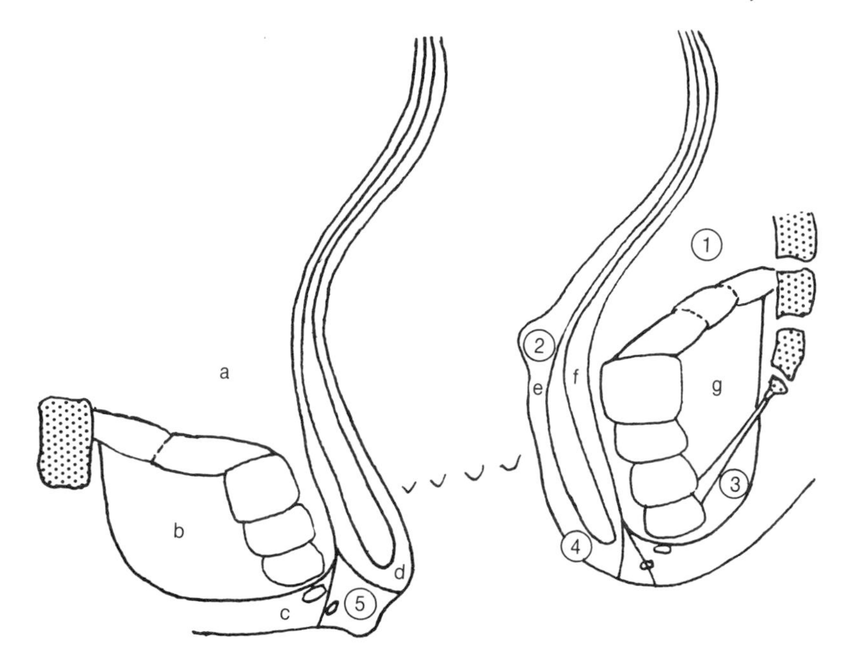

Fig. 4.9 The acute anorectal non-cryptoglandular non-fistulous abscesses of Eisenhammer. Adapted from Eisenhammer S.[9]
(a) Pelvirectal supralevator space.
(b) Ischiorectal space.
(c) Peri-anal or superficial ischiorectal space.
(d) Marginal or mucocutaneous space.
(e) Submucous space.
(f) Anorectal intermuscular space.
(g) Deep postanal space.
(1) Pelvirectal supralevator abscess.
(2) Submucous abscess (infected haemorrhoid injection or trauma).
(3) Ischiorectal foreign body abscess.
(4) Mucocutaneous or marginal abscess (infected haematoma).
(5) Subcutaneous or perianal abscess (follicular skin infection).

referred complicated (and he thought iatrogenic) fistulae. Emergency fistulectomy, in his view, obviates surgery of the chronic fistula-in-ano, a policy supported by others.[12]

Whatever classification is used, it is evident that the majority of fistulae are simple and relatively easy to treat.[7] Classically this means laying open the track, and as this is usually low the sphincters are little damaged, and as a consequence there is little in the way of postoperative functional morbidity. Indeed, the recorded treatment of fistula-in-ano by this method in this country dates back to John of Arderne[13] in the fourteenth century, and it may well be that apart from trephining, operations for fistula were the earliest to be performed.[14]

Fistulae are usually considered more difficult when the primary track crosses the upper part of the external sphincter or when the track is suprasphincteric or extrasphincteric.[15] Another 'difficult' fistula is one that recurs despite skilled attention; overlooked internal openings or secondary tracks usually being responsible for the recurrence. Increasing involvement of the sphincter complex means that conventional laying open by sphincter division will lead to an ever increasing impairment of continence.

Consequences of anal sphincter division

When the anal canal is studied manometrically the reported pressure values in healthy persons may vary according to the method of investigation employed.[16] The internal anal sphincter usually contributes up to 85 per cent to the resting anal pressure; the external anal sphincter increases this pressure when faeces enter the upper part of the canal or upon voluntary contraction. Bennett and Duthie[17] demonstrated a significant reduction in anal pressure after internal anal sphincterotomy, and later, in 1965,[18] found a reduction in anal pressure in the distal 2 cm of the anal canal in patients who had undergone surgery for both high and low fistulae. Belliveau, Thomson and Parks[19] also found reduced resting pressures in the distal 2–3 cm of the anal canal postoperatively compared with preoperatively, although the pressures in patients who had undergone surgery for intersphincteric fistula were not significantly different from controls. Patients with transsphincteric fistulae in whom the external sphincter was divided had significantly lower resting anal pressures than those in whom the muscle was not cut. Voluntary contraction resulted in similar pressures in controls and in patients whose internal sphincters had been laid open, but pressures were significantly lower when the external sphincter had been cut. Patients with transsphincteric or suprasphincteric fistulae in whom the external anal sphincter had been preserved and a loose seton used showed no reduction in contractile pressure (maximum contraction pressure minus resting pressure). There was some association between the incidence of impaired continence and division of the external anal sphincter. Sainio and Husa[20] showed that patients with postoperative functional defects, mainly women, before surgery already had lower resting, squeeze and contractile pressures than patients who subsequently had no functional defect. Further, the degree of reduction in these pressures was greater in women than in men. In a larger, but retrospective study, Sainio[21] demonstrated that internal sphincter division, especially the more extensive division employed in the treatment of transsphincteric fistulae, resulted in the lowest resting anal pressures and the highest incidence of defective anal control.

Although sphincter preservation appears to be the predominant factor in maintaining continence, other factors have been studied. The recto-anal inhibitory reflex has been found by some[18] to be disturbed after fistula surgery, although not by others.[21] Similarly, scar shape and consistency have on occasions been found to be associated with changes in degrees of continence. Thomson and Ross[22] raise the suggestion that disordered anal control after sphincter division may also have a further aetiological component, such as damage to the nerve supply of puborectalis either by sepsis or by surgical dissection.

Attempts to reduce incontinence

How then, may curative surgery be performed with least functional impairment? Parks[11] based his operative treatment on the removal of the supposed diseased anal gland. This means removal of a segment of the internal sphincter, deep to which the anal gland is situated, by internal sphincterectomy; such being deemed a *sine qua non* for effective fistula surgery.[10,12,23]

Preoperative physiology may help the surgeon choose a less radical approach in some cases and thereby enhance continence. Pescatori *et al.*[24] prospectively studied 132 patients divided into two groups: those studied preoperatively manometrically and those who were not. In the former, when there was poor resting tone or voluntary contraction (usually patients over 50 years old and female), internal and external sphincter division was limited to 1 cm; loose setons were used in transsphincteric fistulae. Cases in which preoperative manometry was adequate, a long internal sphincterotomy was performed for high intersphincteric tracks, and transsphincteric fistulae were laid open. All of the second group underwent a 2 cm internal sphincterotomy, followed by laying open low transsphincteric fistulae or using setons in the higher ones. Soiling was significantly reduced with preoperative manometry as well as a lower rate of recurrent fistula. Internal sphincter division led to more incontinence when manometry had been omitted, although the authors felt that some of the change in function may have been because of the simultaneous division of the lowermost fibres of the external sphincter.

Setons

Chemical setons

Wollfers[25] reported the highly successful ayurvedic treatment of fistula-in-ano as used in Columbo. A thread, impregnated 21 times for 15 minutes in a caustic solution prepared from the ashes of herbs, is inserted into the fistula track and changed weekly until healing is complete (up to 12 weeks for 'large' fistulae). In the cases of multiple fistulae, no more than two threads are inserted at any one session. In the series of 80 patients, no recurrences were reported, and the patients, as well as never being formally hospitalized, were able to pursue their employment during treatment. In another Third World report, Misra and Kapur[26] advocate outpatient passage of a braided stainless steel wire, which is tied securely, for all types of fistula in which the external opening is visible. After wiring, analgesics are given liberally and then Sitz baths are given twice daily; antibiotics are used if a purulent discharge appears. The wire is tightened weekly until it has completely cut through. A 4 per cent recurrence rate was reported, and there were no complaints of temporary or permanent incontinence to flatus or faeces.

Kupferberg *et al.*[27] presented a novel approach to the management of patients with recurrent high anorectal fistulae. After formal bowel preparation, patients underwent excision of the fistulous track as high as possible so as not to endanger the sphincter mechanism, closure of the proximal end of the track as close to the rectal wall as possible and insertion of gentamicin-impregnated beads into the wound before complete primary wound closure. From 1 week postoperatively, the chains were gradually extracted over a period of 3 weeks. In their small series of five patients, no recurrences had developed after 2 years of follow-up.

Loose setons

Laying open the primary and secondary tracks is considered the standard treatment of transsphincteric fistulae, resulting in division of a variable amount of external anal sphincter. Although eradication of acute and chronic sepsis is the principal consideration, a good functional result in terms of anal continence should be aimed for. To this end, Thomson and Ross[22] assessed the results of treatment of patients with transsphincteric fistulae in whom the external sphincter was preserved as much as possible with the use of a loose seton. At operation, the primary and secondary tracks were defined and the secondary tracks curetted. The intersphincteric space was laid open by internal sphincterotomy from the anal verge to the level of the primary track. A loose seton was then passed to encircle the primary track in order to:

- Act as a marker of the track
- Act as a drain for the track
- Stimulate fibrosis around it

The loose seton was removed at a later date after examination under anaesthesia had revealed that the rest of the wound had fully healed. The external sphincter was only divided if the primary track failed to heal after seton removal or if there was persistent local sepsis around the seton. Complete healing after seton removal and without recourse to sphincter division occurred in 44 per cent of the patients. When the sphincter was preserved more patients reported full continence than when it was divided, with more patients in the latter group reporting impaired control of flatus, liquid and solid stool.

Kennedy and Zegarra[28] employed a similar technique in 32 patients with high transsphincteric or suprasphincteric fistulae and found recurrence and/or persistence in 34 per cent of posterior fistulae but only in 12 per cent of anterior fistulae.

Tight setons

By contrast to those who advocate a loose seton, Held *et al.*[29] found good functional results and no recurrences when a rubber band seton was used. An acute abscess can first be converted into a chronic draining fistula; and tightening of the band at intervals leads to severance with fibrosis and little separation of the sphincter musculature. Christensen *et al.*[30] assessed functional and manometric results after high transsphincteric fistulae had been treated in this way. Rather than tighten the seton fortnightly it was tightened every second day so that it cut through within 12 days. The authors found that 62 per cent of patients had some degree of incontinence, although manometry was not significantly different between continent and incontinent patients.

Primary closure

Primary closure was attempted by Chaissagnac in 1856 and by Stephen and Smith in 1879, but was given up later because of the fear of serious infection. Rygich,[31] however, obtained a 98 per cent success rate in 1700 cases since 1953; his closure consisting of repair of all divided tissues except the skin. Parkash *et al.*[8] reviewed 120 cases over an 11-year period treated by excision and primary closure. Of the 120 cases, 21 were transsphincteric (one with a high blind infralevator extension) and one suprasphincteric with a high blind supralevator extension. Identification of the tracks was aided where necessary with dye, and stay sutures were left in the muscles to hold and identify the divided structures. The internal opening was cored out and closed by an inverting purse string stitch, unless close to the anal verge in which case it was excised and the mucosal edges repaired. After excision all layers were meticulously repaired (except in some cases the skin), and a subcutaneous lateral sphincterotomy performed for the relief of postoperative pain and promotion of healing. Most patients were given antibiotics according to the microbiological culture of pus obtained at operation. Over 88 per cent healed within 2 weeks. The three recurrences all involved multiple transsphincteric fistulae and were retreated successfully in the same manner; the four cases of wound breakdown were successfully treated by secondary suture or healed by secondary intention.

Lewis[32] advocates an alternative method. The fistula track and its branches are 'cored out' from the external opening inwards, stopping well short of the anal margin. If the fistula is low, then the tunnel left after coring out the fistula is converted into a laid open wound, achieving the same result as cutting down onto a probe but with the following advantages:

- The track is cored out under direct vision
- There is no risk of laying open a false track from incorrect placement of a probe
- Side tracks are not missed as transected granulation tissue is clearly seen
- Blind tracks may be identified and removed without dividing overlying tissue
- The relation of the track to the sphincters is defined before sphincter division
- The entire track is available for histology

If the fistula is high (or of any sort in elderly patients with poor sphincter function) then the tunnel cannot be converted into a laid open wound without some incontinence, so the rectal wall is closed in two layers, followed by closure of the tunnel in the sphincter or pelvic floor. The remaining external wound is lightly packed with ribbon gauze soaked in povidone-iodine. Patients are covered by a 5-day course of antibiotics, and recurrent fistulae are treated in the same way but under the protection of a covering colostomy. Lewis reports only four known recurrences out of over 100 low fistulae, and only one recurrence in 18 transsphincteric and suprasphincteric fistulae.

Advancement flaps (Table 4.3)

Fistulectomy and rectal mucosal advancement is an alternative method of fistula repair that neither involves the use of a seton nor division of the sphincters. Although initially applied to recto-vaginal fistulae,[33] its use in the treatment of non-specific fistula-in-ano is recorded as early as 1912 by Elting.[34] Aquilar *et al.*[35] recommend its use in transsphincteric fistulae claiming that it avoids an associated second procedure when a seton is used, and also reduces the healing time. The fistulous track is excised together with the crypt bearing tissue deep to the internal opening and the overlying anoderm. A layer of mucosa, submucosa and often circular muscle is then mobilized proximally; the internal opening in the muscle is closed with absorbable material; and the mucosal flap is advanced, its own internal opening is trimmed, and the fresh edge is anastomosed to fresh tissue distal to the site of the internal opening. The external wound is left open for drainage. Follow-up of 80 per cent of the 189 patients so treated revealed 10 per cent with minor symptoms of soiling (incontinence to gas or liquid stool) and a 1.5 per cent recurrence rate. Wedell *et al.*[36] have used the sliding flap advancement technique in patients with well-established chronic fistulae classified as suprasphincteric or transsphincteric, with at least half of the external sphincter below the level of the track. In the acute abscess stage, a seton is passed through the main track rather than the track being cored out, and after satisfactory healing (usually about 3 weeks) the advancement flap technique is employed. Reznick and Bailey[37] using the principle of closure of the internal opening, curettage of the track and wide external drainage found excellent results in a small series of patients with recurrent suprasphincteric, high transsphincteric and extrasphincteric fistulae.

Jones *et al.*[38] have used the same technique in patients with recto-vaginal, anorectal and recto-urethral fistulae secondary to inflammatory bowel disease, obstetric trauma and iatrogenic causes, as well as crypto-glandular infection. Interestingly, their success rate and failures did not seem to be affected by the presence or absence of a covering stoma. Lewis and Bartolo[39] recommend its use in fistulae associated with Crohn's disease, reporting no deterioration in continence in the eight patients treated by them using a large full thickness flap of rectal wall.

Table 4.3 The results of 342 endo-anal flap advancements for chronic idiopathic fistula-in-ano show only ten (3.4 per cent) recurrences (the other types, such as those associated with inflammatory bowel disease, have been excluded). This gives a much rosier picture of the procedure than has been the authors' experience

Author	Number	Track type	Flap type	External track	Drainage	Outcome	
						Recurrence	Function
Elting 1912[34]	96	?	Mucosal	Curetted	Pack	1	Stenoses: 7; Grade II: 4
Oh 1983[40]	15	recurrent high	Mucosal	Excised	Drain and pack	2	Satisfactory
Aquilar *et al*. 1985[35]	189	?	Partial/full thickness	Excised	Left open	3	Minor, 10 per cent Grade I: 7 Grade II: 6
Wedell *et al*. 1987[36]	27	suprasphincteric: 4 high-transsphincteric: 12 mid-transsphincteric: 11	Partial	Excised	Pack	0	Unchanged
Jones *et al*. 1987[38]	4	complex	Partial	Excised	Drain	0	?
Shemesh *et al*. 1988[41]	3	anterior high-transsphincteric	Partial	Curetted	Drain	?1	?
Reznick and Bailey 1988[37]	6	transsphincteric suprasphincteric extrasphincteric	Full (3) Direct Closure (3)	Curetted	?	2	Unchanged
Lewis and Bartolo 1990[39]	2	high-transsphincteric	Full	Curetted	Drain	1	Unchanged

Other methods

Another approach is that devised and described by Mann and Clifton[42] in which, by a staged procedure, the extrasphincteric part of a well-established chronic track is re-routed inwards to a site where it may be laid open without any sacrifice of external sphincter muscle, either in the intersphincteric or submucosal planes. After passing a seton through the track and extensively infiltrating the tissue around the track and the inter-sphincteric plane with dilute adrenaline (1:300 000) the track is cored out through the external sphincter or puborectalis. The intersphincteric plane is then developed to the level where the track crosses it, and the external part of the track re-routed into the intersphincteric space, either by passing it through a hole in the external sphincter or puborectalis, or by dividing the muscle. The gap in that divided muscle is then closed. After sound healing of the external wound by secondary intention, the seton can be cut out by division of the lower fibres of the internal sphincter. If the internal opening is high, the track is re-routed further into the submucosal plane, and the submucosal track opened as a third stage.

Conclusions

Fistulae at other intestinal sites will heal if: sepsis or other underlying disease is eradicated; there is no distal obstruction and; the track is not lined by epithelium. By this reckoning, the loose seton technique, for example, should be more effective than it is in fact. There does not seem to be a particular microbiological flora to account for this failure.[43]

For the future, there needs to be a greater attempt to try and unravel the unknown factors relating to fistula development; for example, why the male sex is so much more prone; and why fistulae caused by specific infections will heal with antimicrobial therapy but not the more common variety.

In the meantime, there are good reasons to continue to try and preserve function in patients with anal fistulae. There have have been promising reports of primary repair and advancement flaps, although advancement flaps have been somewhat disappointing in the authors' own experience.

References

1 Saino P. Fistula-in-ano in a defined population. Incidence and epidemiological aspects *Ann Chir Gynaecol*. 1984; **73**: 219–24.
2 Ewerth S, Ahlberg J, Collste G, Holmstrom B. Fistula-in-ano: a six-year follow-up study of 143 operated patients. *Acta Chir Scand*. 1978; **482**: (Suppl) 53.
3 Marks C G, Ritchie J K. Anal fistulas at St Mark's Hospital. *Br J Surg*. 1977; **64**: 84–91.
4 Grace R H, Harper I A, Thompson R G. Anorectal sepsis: microbiology in relation to fistula-in-ano. *Br J Surg*. 1982; **69**: 401–403.

5 Goligher J C, Ellis M, Pissidis A G. A critique of anal glandular infection in the aetiology and treatment of idiopathic anorectal abscesses and fistulas. *Br J Surg*. 1967; **54**: 977–83.
6 Parks A G, Gordon P H, Hardcastle J D. A classification of fistula-in-ano. *Br J Surg*. 1976; **63**: 1–12.
7 Shouler P J, Grimley R P, Keighley M R B, Alexander-Williams J. Fistula-in-ano is usually simple to manage surgically. *Int J Colorect Dis*. 1986; **1**: 113–5.
8 Parkash S, Lakshmiratan V, Gajendran V. Fistula-in-ano: treatment by fistulectomy, primary closure and reconstitution. *Aust N Z J Surg*. 1985; **55**: 23–7.
9 Eisenhammer S. The final evaluation and classification of the surgical treatment of the primary anorectal cryptoglandular intermuscular (intersphincteric) fistulous abscess and fistula. *Dis Colon Rectum*. 1978; **21**: 237–54.
10 Eisenhammer S. Emergency fistulectomy of the acute primary anorectal cryptoglandular intermuscular abscess fistula-in-ano. *S Afr J Surg*. 1985; **23**: 1–7.
11 Parks A G. Pathogenesis and treatment of fistula-in-ano. *Br Med J*. 1961; **1**: 463–9.
12 McElwain J W, Maclean M D, Alexander R M, *et al.*. Anorectal problems: experience with primary fistulectomy for anorectal abscesses. A report of 1,000 cases. *Dis Colon Rectum*. 1975; **18**: 646–9.
13 Arderne J. Treatment of fistula-in-ano, haemorrhoids and clysters. From an early 15th century manuscript translation. Power D (ed). London: Kegan Paul, Trench, Trubner & Co., 1910.
14 Lockhart-Mummery J P. Discussion on fistula-in-ano. *Proc R Soc Med*. 1929; **22**: 1331–41.
15 Seow-Choen, Phillips R K S. Insights gained from the management of problematical anal fistulae at St Mark's Hospital, 1984–88. *Br J Surg*. 1991; **78**: 539–41.
16 Lilius H G. Fistula-in-ano: a clinical study of 150 patients. *Acta Chir Scand*. 1968; **383**: 52–88.
17 Bennett R C, Duthie H L. The functional importance of the internal anal sphincter. *Br J Surg*. 1964; **51**: 355.
18 Bennett R C, Duthie H L. Pressure and sensation in the anal canal after minor anorectal procedures. *Dis Colon Rectum*. 1965; **8**: 131–6.
19 Belliveau P, Thomson J P S, Parks A G. Fistula-in-ano: a manometric study. *Dis Colon Rectum*. 1983; **26**: 152–4.
20 Sainio P, Husa A. A prospective manometric study of the effect of anal fistula surgery on anorectal function. *Acta Chir Scand*. 1985; **151**: 279–88.
21 Sainio P. A manometric study of anorectal function after surgery for anal fistula, with special reference to incontinence. *Acta Chir Scand*. 1985; **151**: 695–700.
22 Thomson J P S, Ross A H McL. Can the external sphincter be preserved in the treatment of transsphincteric fistula-in-ano? *Int J Colorect Dis*. 1989; **4**: 247–50.
23 Schouten W R, van Vroohoven Th. J M V, van Berlo C L J. Primary partial internal sphincterotomy in the treatment of anorectal abscess. *Neth J Surg*. 1987; **39**: 43–5.
24 Pescatori M, Maria G, Anastasio G, Rinallo L. Anal manometry improves the outcome of surgery for fistula-in-ano. *Dis Colon Rectum*. 1989; **32**: 588–92.
25 Wollfers I. Ayurvedic treatment for fistula-in-ano. *Tropical Doctor*. 1986; **16**: 44.

26 Misra M C, Kapur B M L. A new non-operative approach to fistula-in-ano. *Br J Surg*. 1988; **75**: 1093–4.
27 Kupferberg A, Zer M, Rabinson S. The use of PMMA beads in recurrent high anal fistula: a preliminary report. *World J Surg*. 1984; **8**: 970–74.
28 Kennedy H L, Zegarra J P. Fistulotomy without external sphincter division for high anal fistula. *Br J Surg*. 1990; **77**: 898–901.
29 Held D, Khubchandani I, Sheets J, *et al*.. Management of anorectal horseshoe abscess and fistula. *Dis Colon Rectum*. 1986; **29**: 793–7.
30 Christensen A, Nilas L, Christiansen J. Treatment of transsphincteric anal fistulas by the seton technique. *Dis Colon Rectum*. 1986; **29**: 454–5.
31 Rygich A N. Atlas of the operations on the rectum and colon. 1965; p. 80, Moscow. Quoted by Parkash *et al*.
32 Lewis A. Excision of fistula-in-ano. *Int J Colorect Dis*. 1986; **1**: 265–7.
33 Noble G H. New operation for complete laceration of the perineum designed for the purpose of eliminating danger of infection from the rectum. *Trans Am Gynecol Soc*. 1902; **27**: 363.
34 Elting A W. The treatment of fistula-in-ano with especial reference to the Whitehead operation. *Ann Surg*. 1912; **56**: 744–52.
35 Aquilar P S, Plasencia G, Hardy T G, *et al*.. Mucosal advancement in the treatment of fistula-in-ano. *Dis Colon Rectum*. 1985; **28**: 496–8.
36 Wedell J, Meier zu Eissen P, Banzhaf G, Kleine L. Sliding flap advancement for the treatment of high level fistulae. *Br J Surg*. 1987; **74**: 390–91.
37 Reznick R K, Bailey H R. Closure of the internal opening for treatment of complex fistula-in-ano. *Dis Colon Rectum*. 1988; **31**: 116–8.
38 Jones I T, Fazio V W, Jagelman D G. The use of transanal rectal advancement flaps in the management of fistulas involving the anorectum. *Dis Colon Rectum*. 1987; **30**: 919–23.
39 Lewis P, Bartolo D C C. Treatment of transsphincteric fistulae by full thickness advancement flaps. *Br J Surg*. 1990; **77**: 1187–9.
40 Oh C. Management of high recurrent anal fistula. *Surgery*. 1983; **93**: 330–32.
41 Shemesh E I, Kodner I J, Fry R D, Neufeld D M. Endorectal sliding flap repair of complicated anterior anoperineal fistulas. *Dis Colon Rectum*. 1988; **31**: 22–4.
42 Mann C V, Clifton M A. Re-routing of the track for the treatment of high anal and anorectal fistulae. *Br J Surg*. 1985; **72**: 134–7.
43 Seow-Choen F, Hay A J, Heard S, Phillips R S K. Bacteriology of anal fistulae. *Br J Surg*. 1992; **79**: 27–8.

5

Delorme's operation – the St Mark's experience

Asha Senapati, R. John Nicholls, James P.S. Thomson and Robin K.S. Phillips

Introduction

Full thickness rectal prolapse is a distressing and demoralizing condition. Patients are troubled by a protrusion beyond the anal verge, which bleeds and secretes mucus. It is frequently associated with incontinence either because of an underlying weakness in the sphincter mechanism that allows the prolapse to occur, or because the presence of the prolapse protruding through the anal canal leads to poor sphincter function.

That over 100 different procedures have been described to treat this condition,[1] implies that none are entirely satisfactory. Some of these are listed in Table 5.1. In the UK, abdominal rectopexy is the most frequently used operation, usually with the insertion of Ivalon (polyvinyl alcohol) sponge. Although the recurrence rate is low after this procedure, it frequently results in intractable constipation, which can be debilitating. In addition, the operation requires a laparotomy that is best avoided because this condition occurs most frequently in the extremes of life.

The reason for this constipation is difficult to determine. Some have suggested that it is due to scarring and rigidity around the rectum from the implant. Others have suggested that division of the lateral ligaments interferes with the nerve supply to the rectum. Others blame the redundant sigmoid and recommend a sigmoid colectomy instead,[2] in spite of a 20 per cent unsatisfactory result.

In a series of 59 patients treated by extended abdominal rectopexy[3] (suspension of the uterus and re-attachment of the lateral ligaments to the sacral promontory, in addition to the convential mobilization for an Ivalon rectopexy) there was no mortality and a 12 per cent morbidity. There were no recurrences, but the percentage of constipated patients rose from 29 per cent before surgery to 47 per cent after. There was a 38 per cent improvement in incontinence.

151 cases treated by Ivalon rectopexy alone[4] reported a 2.6 per cent mortality and a 3 per cent recurrence rate all within 3 years. 27 per cent of their patients were constipated after surgery but it is not clear how many

Table 5.1 Some of the procedures performed for the treatment of rectal prolapse

Procedure
Abdominal
Ivalon sponge rectopexy
Ripstein teflon sling rectopexy
Anterior resection and rectopexy
Low anterior resection
Extended abdominal rectopexy
Marlex mesh rectopexy
Roscoe–Graham repair
Loygue rectopexy
Puborectalis sling
Rectal plication
Hartmann's operation
Retroperitoneal colopexy
Absorbable rectopexy
Abdomino-perineal resection
Presacral suture rectopexy
Ivalon stent
Lahaut's operation
Moschcowitz operation
Devadhar's operation
Perineal
Thiersch wire
Silastic sling
Perineal rectopexy
Intersphincteric Ivalon rectopexy and post-anal repair
Post-anal repair
Perineal proctectomy, posterior rectopexy and post-anal repair
Trans-anal fixation
Modified Thiersch procedure
Angelchick prosthesis insertion
Graciloplasty
Rectosigmoidectomy
Altemeier procedure
Alum injection
Delorme's operation

of these suffered this symptom preoperatively. 52 per cent of the patients had improvement in continence.

In another study of 40 patients treated by Ivalon rectopexy[5] there were no deaths and a 10 per cent recurrence rate. Constipation is not discussed in detail but there was a 71 per cent improvement in continence.

In a large series of 101 patients treated by Ivalon rectopexy at St Mark's Hospital[6] no mortality, a 6 per cent morbidity and a 3 per cent recurrence rate resulted. 73 per cent had improvement in continence but 29 per cent had constipation after surgery having not had this symptom before. 9 per cent also had urinary function disturbances.

Boulos *et al.*[7] reviewed 40 young patients having this operation and reported a 20 per cent recurrence rate, three of which recurred after 5 years.

There was a 55 per cent improvement in continence but none of the patients became constipated as a result of the surgery.

Rogers and Jeffery[8] who treated 24 patients with Ivalon rectopexy and a post-anal repair reported a 4 per cent recurrence rate and a 100 per cent improvement in continence. They do not discuss constipation.

Others have reported results after a Marlex mesh rectopexy[9] in 106 patients. There were no recurrences with 86 per cent of the patients being followed-up for over 2 years. However, constipation was not discussed.

Rectopexy combined with anterior resection is popular in the USA. In a series of 102 patients[2] of whom 81 per cent were followed-up for more than 2 years, 2 per cent have developed recurrences. 20 per cent, however, considered their operation to be unsatisfactory either because of constipation or incontinence.

The Ripstein operation (anterior rectopexy) was performed in 108 patients[10] with a 3 per cent mortality and 4 per cent recurrence rate. Defaecation disturbances, however, increased from 27 per cent to 43 per cent.

Estimating the prevalence of constipation after surgery is distorted, if overall prevalance is considered rather than determining whether patients with normal preoperative function have deteriorated. This was done prospectively by Madden *et al.*[11] who found that 42 per cent of patients who were unaffected before Marlex rectopexy became constipated afterwards, using strict criteria for the definition of constipation. Broden *et al.*[12] also found that 40 per cent of patients who were not constipated before a Ripstein operation became so after surgery.

Urinary problems have been reported after rectopexy. Retention postoperatively was seen in 27 per cent in one series.[3]

Pelvic sepsis after an implantation rectopexy is an additional complication of the procedure. It occurs in between 0 and 2.6 per cent of cases[3,4,6,9] and can be troublesome. Lake *et al.*[13] recommend laparotomy for removal of the infected material whereas Ross and Thomson[14] suggest extraction per rectum.

The Delorme's operation, a modification of the operation described by Edmund Delorme in 1900,[15] is a perineal procedure and is not associated with postoperative constipation. This and the additional advantage that, unlike a rectopexy, it can be fairly easily repeated should it fail, have caused its increasing popularity. This led to its use in selected patients in whom the drawbacks of an abdominal procedure were undesirable.

Patients

Between the years 1978 and 1990, Delorme's operation was performed on 32 selected patients at St Mark's Hospital (Fig. 5.1). Their mean age was 69.6 years ± 19.9 (SD) with a range from 14–92 years. There were 24 women and eight men. The duration of their symptoms ranged from 2 weeks to 10 years. 28 (88 per cent) were incontinent preoperatively. 10 (31 per cent) were constipated preoperatively. 13 patients (41 per cent) had had previous surgery for prolapse (Table 5.2) some more than one, making a total of 21 previous operations.

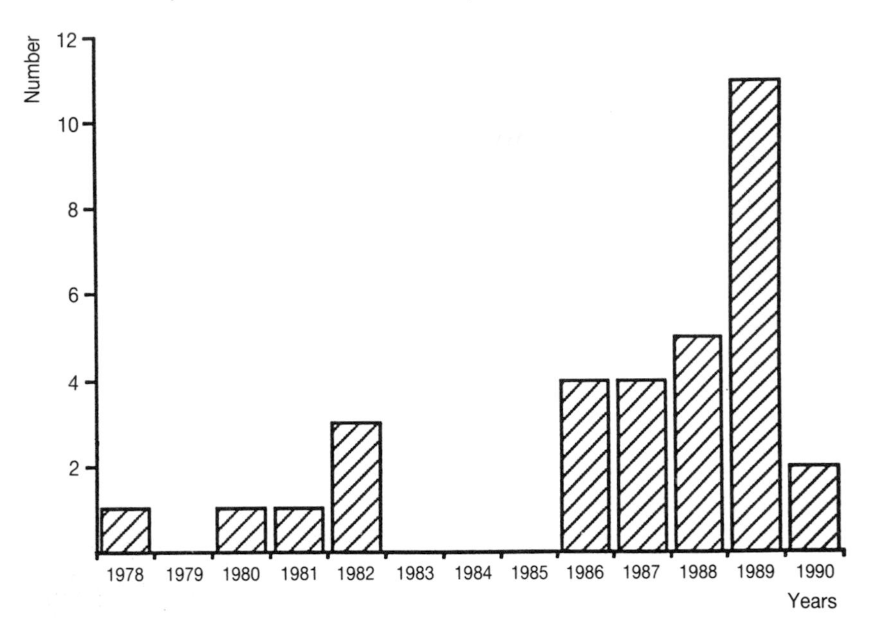

Fig. 5.1 Distribution of Delorme's operations performed at St Mark's Hospital.

Table 5.2 List of previous operations (13 patients: six had two operations; one had three operations)

Ivalon rectopexy	9
Thiersch wire	5
Marlex rectopexy	1
Ripstein rectopexy	1
Anterior resection	1
Postanal repair	1
Excision mucosal prolapse	1
Haemorrhoidectomy	1
Delorme's operation	1
21 procedures	

Operative technique

This is illustrated in the operative photographs and diagrams in Figs 5.2–5.5. The rectum is prolapsed to its full extent and a circumferential

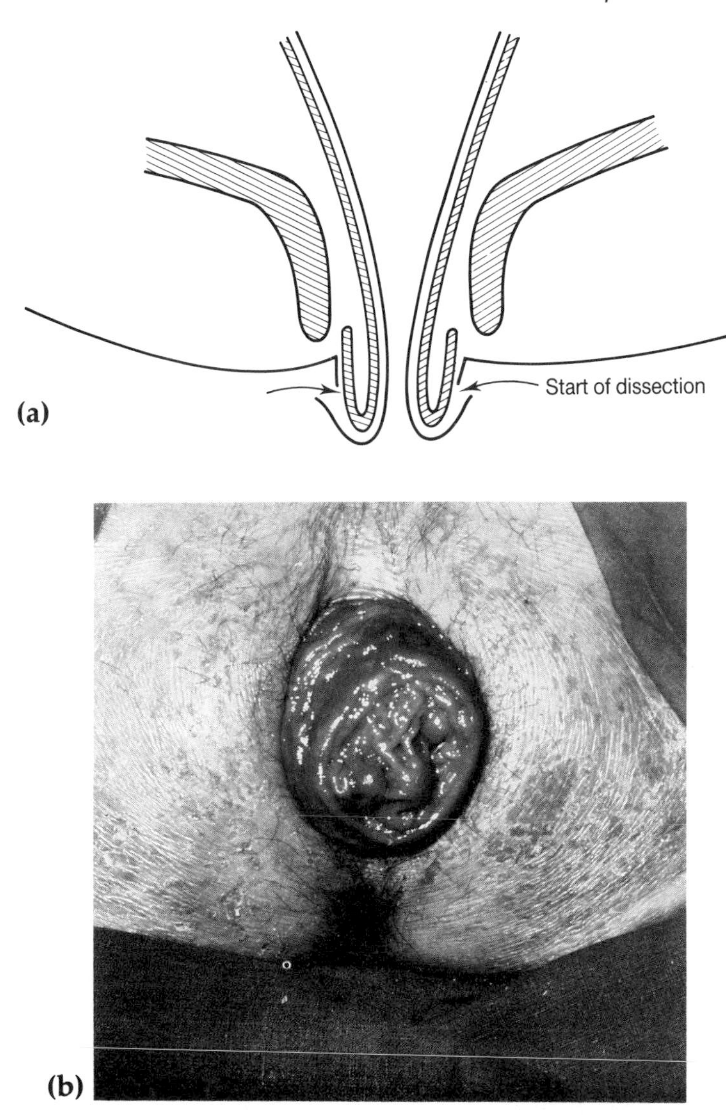

Fig. 5.2 Full thickness rectal prolapse.

incision is made in the mucosa 2 cm above the dentate line. The mucosa is then dissected off the underlying muscle. This dissection may be facilitated either by repeated submucus injections of dilute adrenalin, or by using diathermy, or both. The mucosa is separated circumferentially. As the dissection proceeds first over the external aspect and then the internal aspect of the prolapse, the surgeon is left holding a cylinder of mucosa in his or her non-dominant hand, which allows traction. The dissection proceeds until no further rectal wall can be prolapsed by traction. The

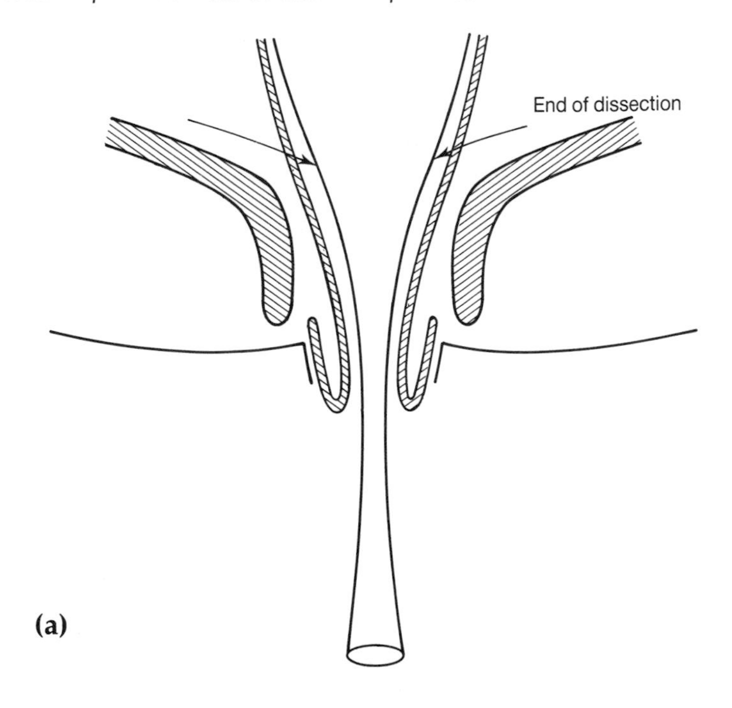

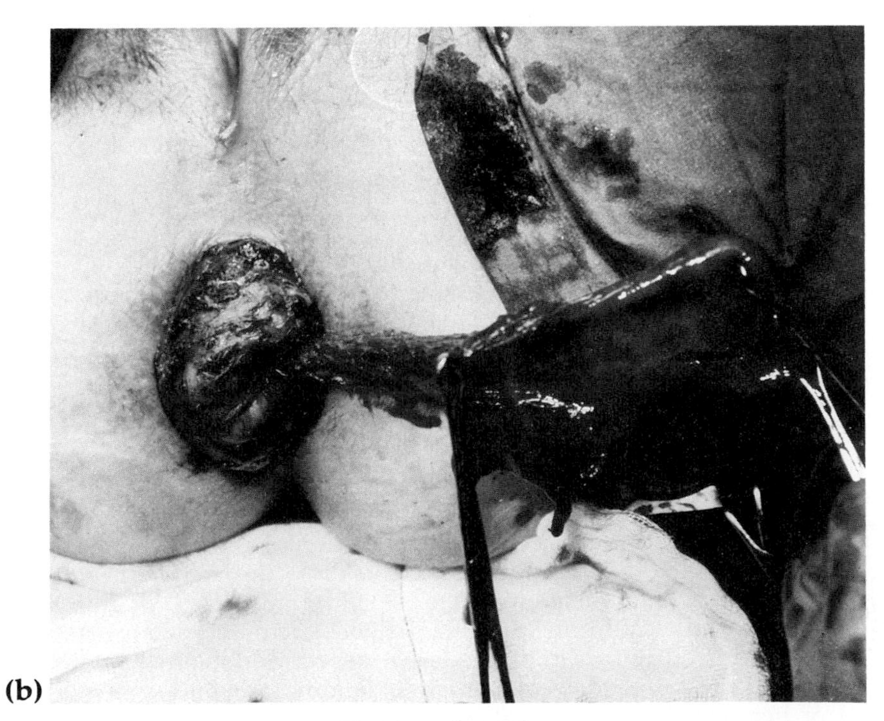

Fig. 5.3 Excision of rectal mucosa.

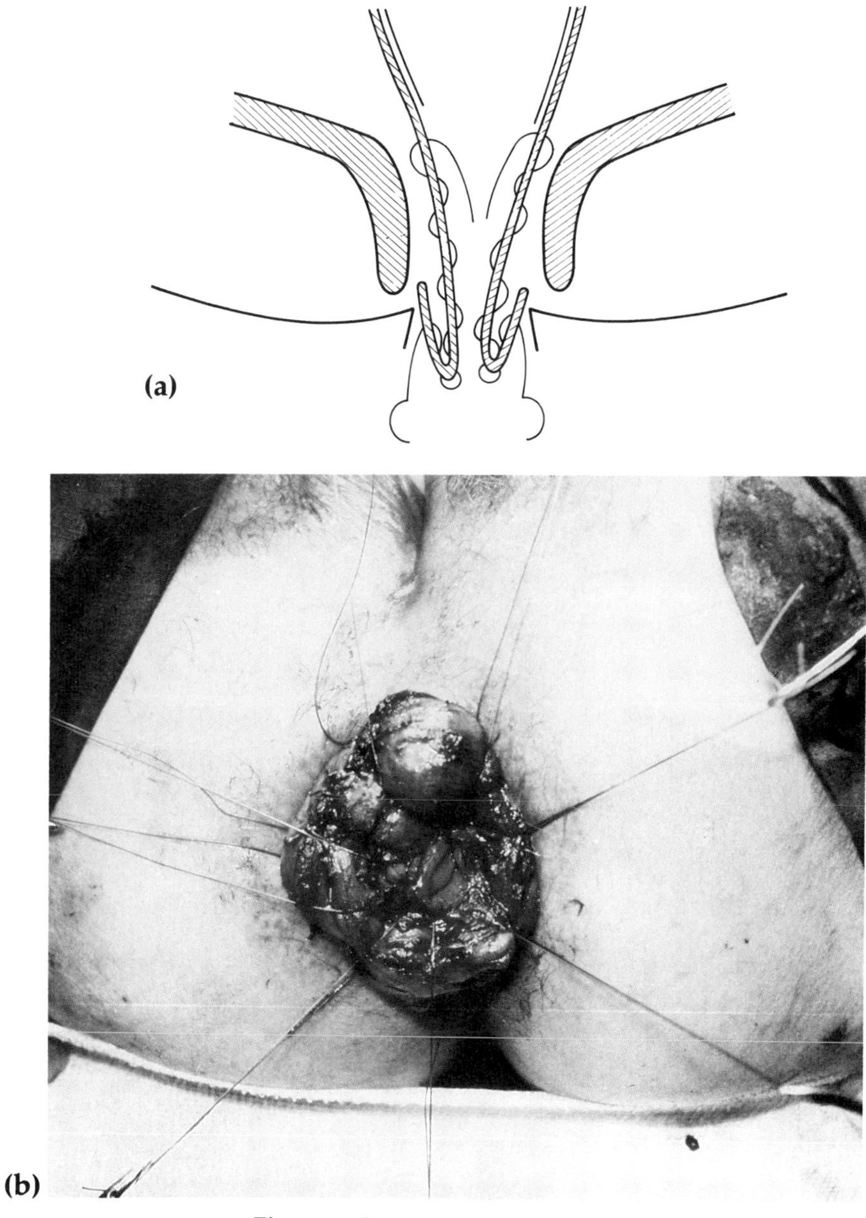

Fig. 5.4 Insertion of plication sutures.

cylinder of mucosa becomes narrow and may be up to 40 cm in length. If a full thickness defect is made in the bowel wall, recognized by seeing fat, this is repaired immediately, but is not a cause for concern as it will be taken up in the plication. Several (between four and eight) plicating absorbable (e.g. Vicryl) sutures are then inserted into the muscle starting from the top of the dissection and finishing just above the dentate line, but not including the mucosa. These are then tightened drawing the proximal dissection margin down towards the distal margin. The mucosal cylinder is then divided, and mucosa to mucosa interrupted sutures inserted to complete the anastomosis.

A modification of this procedure,[16] not performed in any of this series, is to tighten the plicated rectal wall posteriorly in order to tighten the sphincter mechanism.

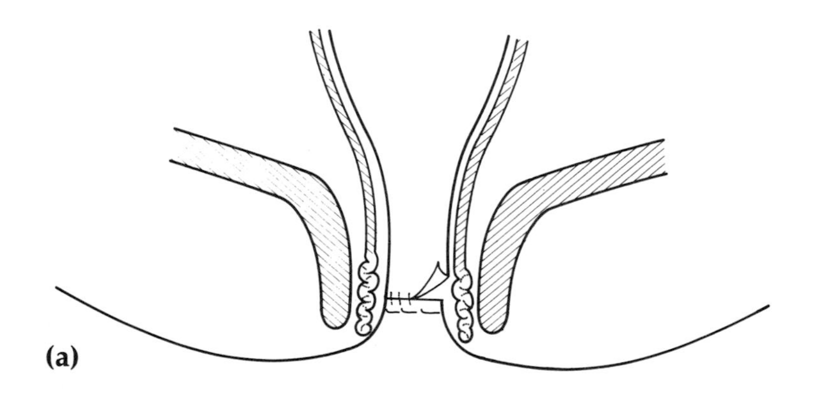

(a)

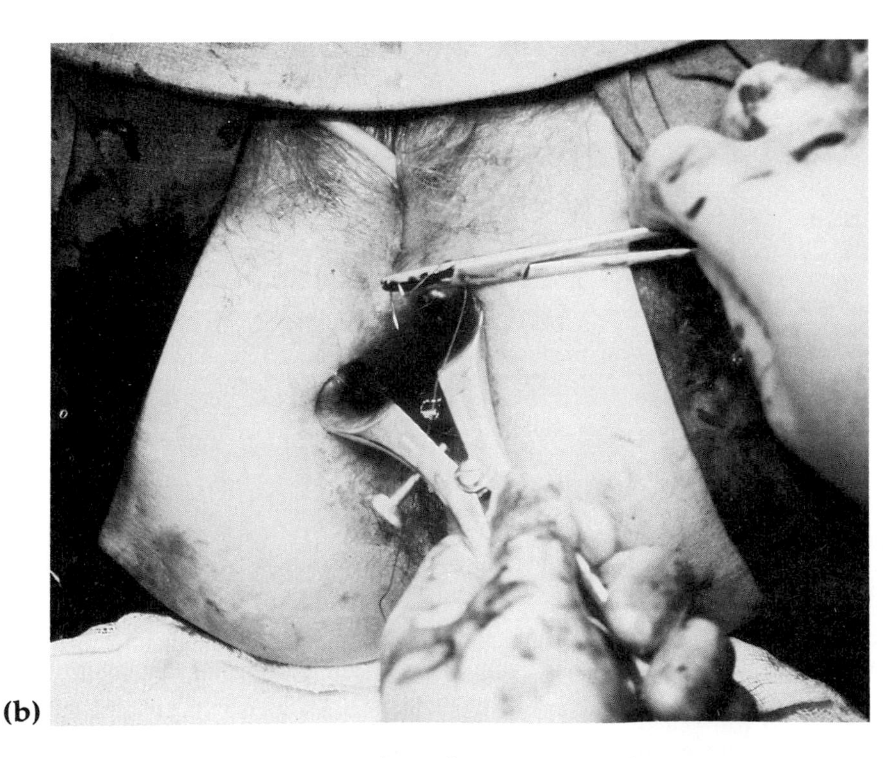

(b)

Fig. 5.5 Plication of muscle and endo-anal anastomosis.

Results

Follow-up has been complete in all 32 patients. Nine patients died from unrelated causes, from 1 month to 10 years later. They were followed up until death. The remainder have been followed from 4 months to 8 years (mean 21.4 months). The mean duration of the operation was 64.6 minutes ± 20.2 (SD), ranging from 40–120 minutes. No patients required a blood transfusion.

There were no deaths related to the operation. Two patients (6 per cent) had complications. One had a wound dehiscence and one had a chest infection. In both, their discharge from hospital was delayed. The median postoperative stay was 10 days.

Four patients had a full thickness recurrence (12.5 per cent). One patient had a further Delorme's operation. One had an Ivalon rectopexy and one is awaiting rectopexy. One patient died before further treatment could be given.

Three patients (9 per cent) have developed a mucosal prolapse postoperatively. One has been injected and two have not required treatment.

Of the ten patients with constipation before surgery, five (50 per cent) improved and the remainder were unchanged. Constipation did not develop in any of the 22 who were not constipated preoperatively. Of the 28 patients with incontinence before surgery, 13 (46 per cent) improved after the operation.

Discussion

Delorme's description of this operation[15] did not involve plication of the muscle. He described three patients, one of whom died of sepsis. Interest waned in the technique for many years, but since the 1970s a handful of papers have been published describing this operation. Their results[16–22] are summarized in Table 5.3.[3] No deaths have been reported. The complication rate is low when stated. The recurrence rate varies from 5–20 per cent. The length of follow-up is usually 2 years or more.

The problem of postoperative constipation after Delorme's operation is not addressed in these papers but rectal stenosis has been reported in up to 9 per cent of patients in two series.[20,22] In our series there were none.

Urinary problems were not encountered in our series perhaps because our patients were routinely catheterized perioperatively. No long-term urinary dysfunction was noted on review of our patients. However, an incidence of 23 per cent urinary retention was reported in one series.[16]

The results of Delorme's operation are rightly judged in the context of the more commonly practiced procedures, and although the recurrence rate after the operation is undoubtedly higher than after abdominal procedures there is the advantage of avoiding an abdominal operation allowing patients in some situations to be treated as day cases. It also appears to avoid the hazard of intractable constipation seen after implantation rectopexy.

Table 5.3 Summary of publications on the Delorme's operation

Author	Year	Number of patients	Recurrence	Postoperative constipation	Improvement in continence	Follow-up	Urinary dysfunction	Other complications	Mortality
Uhlig[16]	1979	44	3(6.8%)	NS	NS	2–10 years	10(23%)	5(11%)	0
Christiansen[17]	1981	12	2(17%)	NS	50%	median 3 years	NS	0	0
Gunderson[18]	1985	18	1(6%)	NS	NS	mean 42 months 3 months–9 years	NS	3(17%)	0
Monson[19]	1986	27	2(7.4%)	NS	83%	mean 35 months 11–64 months	NS	0	0
Houry[20]	1987	18	3(17%)	NS	44%	mean 18 months 8 months–4 years	NS	NS	0
Heaton[21]	1988	5	1(20%)	NS	NS	mean 11 months	NS	0	0
Abulafi[22]	1990	22	1(5%)	NS	75%	mean 29 months 3–70 months	3(14%)	3(14%)	0

NS = not stated

Further surgery after Delorme's operation seems to be for the treatment of recurrence. Although the recurrence rate is low after abdominal procedures, additional surgery is needed in up to 7 per cent[3,10,11] of patients for complications unrelated to recurrent prolapse.

Incontinence is improved after all procedures for the treatment of rectal prolapse. Part of the reason for this may be the cessation of the mucus leakage from the prolapse previously interpreted as incontinence. Most studies are retrospective and have not classified and recorded incontinence correctly. Nonetheless, there is undoubtedly some real improvement. The reasons for this are said to be due to the tone of the internal and external sphincter being allowed to recover when the prolapse has been treated. Resting anal pressure has been shown to increase after a Ripstein procedure whether or not continence was restored.[23] Broden *et al.*[24] also found a significant increase in resting pressure but not squeeze pressure after surgery. Yoshioka *et al.*[25] found significantly lower anal canal pressure levels in patients with rectal prolapse compared to controls. However, none of these values rose significantly after surgery in spite of continence being restored in the majority. In an earlier publication from the same department[26] reduction in basal and squeeze pressures were found only in those patients with prolapse and incontinence but not in those with prolapse alone. There was no significant rise in pressures after surgery and these were not related to the clinical results. In another study[27] prolapse was shown to be associated with slow transit constipation suggesting an aetiological factor.

The results of the selected series presented in this chapter are in keeping with the published literature, in spite of a third of the patients having had previous surgery. In unselected patients, therefore, one would perhaps expect the results to be better still. This allows us to recommend that this procedure be considered in the first instance for patients with full thickness rectal prolapse.

References

1 Wassef R, Rothenberger D A, Goldberg S M. Rectal prolapse. *Curr Prob Surg*. 1986; **23**: 398–451.
2 Watts J D, Rothenberger D A, Buls J G, *et al.*. The management of procidentia. *Dis Colon Rectum*. 1985; **28**: 96–102.
3 Mann C V, Hoffman C. Complete rectal prolapse: the anatomical and functional results of treatment by an extended abdominal rectopexy. *Br J Surg*. 1988; **75**: 34–7.
4 Morgan C N, Porter N H, Klugman D J. Ivalon (polyvinyl alcohol) sponge in the repair of complete rectal prolapse. *Br J Surg*. 1972; **59**: 841–6.
5 Atkinson K G, Taylor D C. Wells procedure for complete rectal prolapse. *Dis Colon Rectum*. 1984; **27**: 96–8.
6 Penfold J C B, Hawley P R. Experiences of Ivalon – sponge implant for complete rectal prolapse at St Mark's Hospital 1960–1970. *Br J Surg*. 1972; **59**: 846–8.
7 Boulos P B, Stryker S J, Nicholls R J. The long-term results of polyvinyl alcohol (Ivalon) sponge for rectal prolapse in young patients. *Br J Surg*. 1984; **71**: 213–4.

8 Rogers J, Jeffery P J. Postanal repair and intersphincteric Ivalon sponge rectopexy for the treatment of rectal prolapse. *Br J Surg*. 1987; **74**: 384–6.
9 Keighley M R B, Fielding J W L, Alexander-Williams J. Results of Marlex mesh abdominal rectopexy for rectal prolapse in 100 consecutive patients. *Br J Surg*. 1983; **70**: 229–32.
10 Holmstrom B, Broden G, Dolk A. Results of the Ripstein operation in the treatment of rectal prolapse and internal rectal procidentia. *Dis Colon Rectum*. 1986; **29**: 845–8.
11 Madden M V, Kamm M A, Nicholls R J, *et al.*. Abdominal rectopexy for complete rectal prolapse: a prospective study evaluating changes in symptoms and anorectal function. *Dis Colon Rectum*. 1992; **35**: 48–55.
12 Broden G, Dolk A, Holmstrom B. Evacuation difficulties and other characteristics of rectal function associated with procidentia and the Ripstein operation. *Dis Colon Rectum*. 1988; **31**: 283–6.
13 Lake S P, Hancock B D, Lewis A A M. Management of pelvic sepsis after Ivalon rextopexy. *Dis Colon Rectum*. 1984; **27**: 589–90.
14 Ross A H McL, Thomson J P S. Management of infection after prosthetic abdominal rectopexy (Wells' procedure). *Br J Surg*. 1989; **76**: 610–2.
15 Delorme E. On the treatment of total prolapse of the rectum by excision of the rectal mucous membranes. *Bulletin et Memories de la Societe de Chirurgie de Paris*. 1900; **26**: 499–518.
16 Uhlig B E, Sullivan E S. The modified Delorme operation. *Dis Colon Rectum*. 1979; **22**: 513–21.
17 Christiansen J, Kirkegaard P. Delorme's operation for complete rectal prolapse. *Br J Surg*. 1981; **68**: 537–8.
18 Gundersen A L, Cogbill T H, Landercasper J. Reappraisal of Delorme's procedure for rectal prolapse. *Dis Colon Rectum*. 1985; **28**: 721–4.
19 Monson J R T, Jones N A G, Vowden P, Brennan T G. Delorme's operation: the first choice in complete rectal prolapse? *Ann R Coll Surg Engl*. 1986; **68**: 143–6.
20 Houry S, Lechaux J P, Huguier M, Molkhou J M. Treatment of rectal prolapse by Delorme's operation. *Int J Colorect Dis*. 1987; **2**: 149–52.
21 Heaton N D, Rennie J A. Extended abdominal rectopexy. *Br J Surg*. 1988; **75**: 828.
22 Abulafi A M, Sherman I W, Fiddian R V, Rothwell-Jackson R L. Delorme's operation for rectal prolapse. *Ann R Coll Surg Engl*. 1990; **72**: 382–5.
23 Holmstrom B, Broden G, Dolk A, Frenckner B. Increased anal resting pressure following the Ripstein operation. *Dis Colon Rectum*. 1986; **29**: 485–7.
24 Broden G, Dolk A, Holmstrom B. Recovery of the internal anal sphincter following rectopexy: a possible explanation for continence improvement. *Int J Colorect Dis*. 1988; **3**: 23–8.
25 Yoshioka K, Hyland G, Keighley M R B. Anorectal function after abdominal rectopexy: parameters of predictive value in identifying return of continence. *Br J Surg*. 1989; **76**: 64–8.
26 Keighley M R B, Makuria T, Alexander-Williams J, Arabi Y. Clinical and manometric evaluation of rectal prolapse and incontinence. *Br J Surg*. 1980; **67**: 54–6.
27 Keighley M R B, Shouler P J. Abnormalities of colonic function in patients with rectal prolapse and faecal incontinence. *Br J Surg*. 1984; **71**: 892–5.

6

Clinical factors which influence outcome after colorectal cancer surgery

Robin K.S. Phillips

Introduction

When a person presents with colorectal cancer the disease is either truly local or has already spread elsewhere. It is not always obvious that metastases are established, even when the most sophisticated methods of imaging are employed. If all patients with a diagnosis of colorectal carcinoma are considered, only about 55 per cent will be suitable for a clinically curative operation and afterwards leave hospital alive, whereas the rest will have presented with malignant ascites, inoperable primary tumours or removable primary tumours but obvious distant metastases. Of the 55 per cent who leave hospital alive after a curative operation, roughly half will die of cancer during the next 5 years. The other half can be considered cured.[1] Presumably those that die of cancer represent a group with occult hepatic metastases;[2] although some might argue that a small proportion of these patients might have been cured if a policy of early vascular ligation had been carried out or if blood transfusion had been avoided.

It follows that there are outcomes that can be influenced by the surgeon – postoperative death, local recurrence and function – and outcomes that are already in-built at presentation. The most important of these latter outcomes is long-term survival, but the condition of the patient at presentation, and the locally advanced nature of the tumour in a number of patients, sometimes leads to unavoidable mortality in hospital despite every clinical skill and effort.

Death in hospital

Death in hospital is largely determined by clinical factors, such as the patients' age and general health, urgent presentation with either obstruction or perforation, and the surgeon's skill.[3] The 'Confidential Enquiry into Perioperative Deaths' (CEPOD) studied the 30-day mortality from three National Health Service regions.[4] Of 4034 deaths, 2391 (59 per cent) were evaluated. The most common diagnoses in patients who died were: 12 per cent fractured neck of femur; 7 per cent intestinal obstruction; 5 per cent cancer of the colon; 5 per cent peptic ulcer; and 5 per cent aortic aneurysm. 79 per cent of all deaths occurred in people who were over the age of 65 years, despite the fact that only 22 per cent of all operations are performed in this age group. In itself this may not seem surprising, but the report clearly showed that a percentage of these deaths might have been avoided if the more senior surgeons and anaesthetists had been more involved in the management of emergency cases; if more difficult cases had been referred to specialists rather than the operations being performed by general surgeons; by better training; and by less hasty decisions to operate when the patient's condition was unstable.[5]

Tumour site

The site of a colorectal cancer plays some part in the risk of hospital death. In the 'Large Bowel Cancer Project' the overall death rate was 11 per cent but was statistically significantly different for various sites within the colon and rectum (right colon 12 per cent; splenic flexure 18 per cent; left colon 12 per cent; rectosigmoid 11 per cent; rectum 10 per cent). The high risk at the splenic flexure was mainly accounted for by deaths after emergency surgery.[6]

The patient's age

Lewis and Khoury[7] reported on 277 consecutive patients over the age of 70 who underwent surgery for colorectal cancer between 1975 and 1985. They found that the overall mortality was 11 per cent, but after palliative resection the mortality was much higher (28 per cent). A similar study from Switzerland of 140 patients with gastrointestinal cancer, who were all aged over 80, found the mortality rate to be 17 per cent.[8] In the Large Bowel Cancer Project the mortality for patients less than 70 years old was 3 per cent, but 12 per cent in those older than 70 years.[3]

The influence of obstruction or perforation

Emergency colorectal cancer surgery in old patients is followed by a much higher mortality. In a study[9] of 545 patients undergoing either elective or emergency surgery, a mortality rate of 38 per cent was found when emergency surgery was performed in patients aged over 80 years, compared with 6 per cent in patients aged less than 75 years and 24 per cent in those between 75 and 80 years. This same trend was seen by the Large Bowel Cancer Project: when obstruction was present the mortality was 32 per cent and 21 per cent for patients aged 80+ and 70–79 years respectively; when obstruction was absent the rates were 16 per cent (80+ years) and 8 per cent (70–79 years).[3]

The influence of the surgeon

Mortality in emergency cases reflects also the seniority of the surgeon involved. The Large Bowel Cancer Project reported a 24 per cent mortality for primary resection in the presence of intestinal obstruction when the operation was performed by surgeons in training, which compared unfavourably with the 13 per cent achieved by Consultants.[10] Other factors besides are probably just as important, however, for example, it is also usual practice for a senior anaesthetist and a more skilled assistant surgeon to be present when a Consultant surgeon is doing the operation.

Even between individual Consultant surgeons surgical skill plays a part. In the Large Bowel Cancer Project the clinical factors with a statistically significant effect on mortality were: age; presentation; obstruction; perforation; tumour fixity; intra-abdominal sepsis; cardiopulmonary complications; anastomotic leak. Some surgeons had a leak rate as low as 0.5 per cent whereas others had a leak rate as high as 30 per cent.[11] When there was no anastomotic leak the mortality was 4 per cent, but when a leak occurred (as it did in 11 per cent of cases) the death rate rose to 19 per cent.[3]

Local recurrence and survival

Local recurrence rates vary widely in the literature. For example, when an aggressive policy towards the management of more disseminated disease is taken, as was the case in the USA with Wangensteen's second look laparotomy policy[12] and remains the case there because of a greater use of both systemic chemotherapy and implantable hepatic perfusion devices, some quite remarkably high rates of local recurrence are reported (of the order of 30–40 per cent).[13,14] Similarly, series where the post-mortem rate is high, for example, Malmo in Sweden, also describe high rates of local failure.[15] These rates must be considered to represent the biological problem of local recurrence. By contrast, in the UK local recurrence rates of the order of 10–15 per cent are accepted as the norm.[16] Such low rates probably describe the clinical problem of local recurrence; where there is no clinical problem it is not usual to submit the patient to extensive tests in the UK – a patient with multiple liver metastases will often not be treated aggressively and, therefore, is not further investigated.

Causes of local recurrence

Local recurrence may occur because: the primary tumour has been disrupted; local excision has been inadequate; viable exfoliated cancer cells have been introduced into the anastomosis or the wound; or because a second primary tumour arises at the anastomosis.

Clinical factors that are associated with a higher than usual chance of local recurrence are: the presence of obstruction or local perforation; the site of the primary tumour; the surgeon who performs the operation; and perhaps the choice of operation.[16] Pathological factors such as the stage and (possibly) grade of the tumour are not considered further.

The site of the tumour

Low rectal tumours have a high rate of local recurrence. For example, Gilchrist and David[17] found 16 per cent local recurrences in 112 patients undergoing abdominoperineal excision for tumours that were situated partly or wholly below the peritoneal reflection as opposed to 3.6 per cent in 55 patients with intraperitoneal growths. Stearns and Binkley[18] also found that the frequency of local recurrence was inversely proportional to the height of the tumour from the anal verge. They studied 369 patients followed-up after abdominoperineal excision of the rectum performed between 1928 and 1945. A local recurrence occurred in 30 per cent of patients whose primary tumour had been situated in the lower third of the rectum, 20 per cent in the middle third and 14.5 per cent in the upper third. This pattern has been noted by many authors reporting results either after abdominoperineal excision or anterior resection[19–22] and the topic has been extensively reviewed by Goligher.[23] More recently, Theile *et al.*[24] have analysed the pattern of tumour recurrence from a prospective study of 210 patients who underwent curative surgery for rectal cancer at the Princess Alexandra Hospital, Brisbane, Australia between 1971–80. More pelvic recurrences developed after surgery for distal tumours: lower third 16 per cent; middle third 13 per cent; upper third 9.5 per cent.

Elsewhere in the colon there is some disagreement whether the **survival** of patients with right-sided tumours is better[25,26] or worse[27] than left. Slaney[1], who studied 12 494 cases of colorectal cancer reported to the Birmingham Regional Cancer Registry between 1950 and 1961, found no difference in ultimate prognosis between right- and left-sided tumours (approx. 40–45 per cent 5-year survivors for radical surgery), but noted a rather worse survival for splenic flexure (30 per cent). **Local recurrence** rates have been compared rarely. Olson and his colleagues,[28] who reported on 281 patients, found no regional recurrences in 34 whose primary tumour was situated either in the transverse or left colon. In the right colon, 95 patients were studied with 5 per cent and in the sigmoid 85 patients with 9 per cent local recurrences.

The Large Bowel Cancer Project found that patients with a splenic flexure carcinoma had a significantly reduced age-adjusted 5-year survival when compared with other sites within the colon (right colon 70 per cent; left colon 64 per cent; splenic flexure 50 per cent). Although splenic flexure carcinomas are more likely to obstruct, and although obstructing tumours have a worse survival than non-obstructing tumours, neither obstruction nor the patient's age, sex, Dukes' stage and tumour differentiation could account for the worse survival.[6] The chance of **local recurrence** was also high at the splenic flexure (right colon 15 per cent, splenic flexure 23 per cent, left colon 14 per cent, rectosigmoid 15 per cent). This finding could not be explained in pathological terms (e.g., Dukes' C: right colon 28 per cent, splenic flexure 29 per cent, left colon 26 per cent, rectosigmoid 27 per cent).[6] The most likely explanation is embryological. The splenic flexure is a watershed between the midgut and hindgut. There are inconsistencies in colonic vascular anatomy so that in 6 per cent of post-mortem dissections there is no left colic artery, its function being taken over by an enlarged middle colic artery; and in 22 per cent the middle colic artery is absent and the right colic artery joins the marginal at the hepatic flexure.[29] All this means that it is often not possible to predict which route the lymphatic drainage from the splenic flexure will take. As cancer surgery should encompass the possible lymphatic drainage, and as the lymphatic drainage follows the arterial blood supply, the right colic, middle colic and left colic arteries should be removed from their origins and the caecum or ileum anastomosed to the sigmoid colon in cases of splenic flexure carcinoma. In addition, there may be a case for removal of the spleen with its hilar lymph nodes, particularly if the tumour is locally advanced.

Obstruction and perforation

Obstructing and perforated tumours are well known to carry a high risk of subsequent local recurrence, a high in-hospital mortality, and a poor survival.[23,30] Ohman reported a retrospective analysis from the Karolinska Hospital in Sweden of 148 patients with malignant large bowel obstruction. These constituted 14 per cent of 1072 colorectal cancer patients seen by the Department of Surgery between 1950 and 1979. The overall survival was 16 per cent in patients with obstruction, less than half the rate for non-obstructed patients. A similarly bad survival rate for obstructing tumours was seen in the Large Bowel Cancer Project (age-adjusted 5-year survival: obstructed 25 per cent; not obstructed 45 per cent). Most of the difference in survival was seen in the first 18 months after surgery after which time the survival curves became parallel, which suggests a similar death rate (Fig. 6.1). It is likely that the survival disadvantage for obstructing tumours

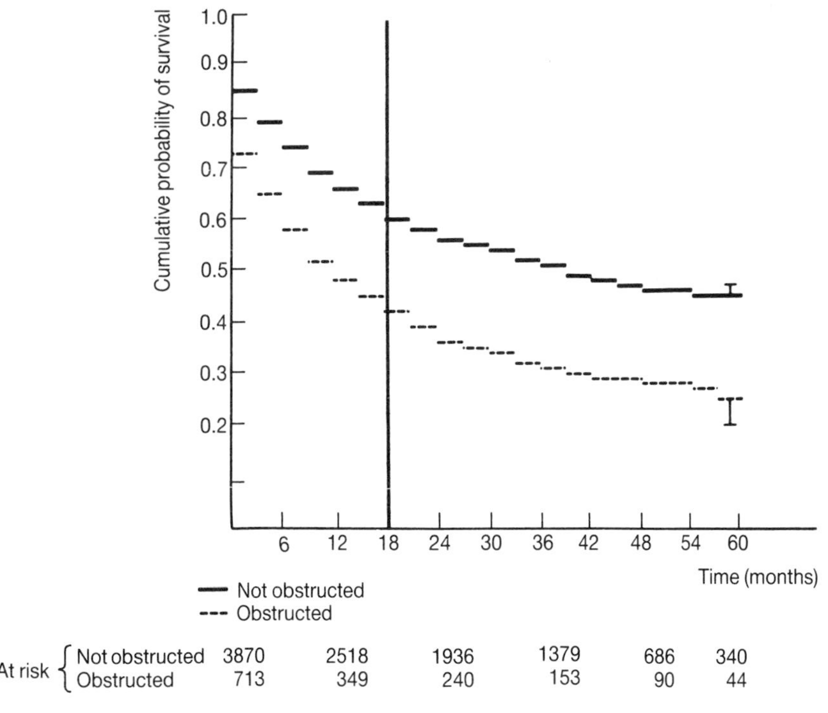

Fig. 6.1 The poor survival after surgery for malignant large bowel obstruction is largely accounted for by a high in-hospital mortality. The survival curves become parallel after the first 18 months, which indicates that the death rate from then on is similar. This supports the view that **pathological** factors (which might exert their influence beyond 18 months) are unlikely to account for the difference in outcome between obstructed and non-obstructed cases. Reproduced by permission of the publishers, Butterworth-Heineman Ltd, from Phillips *et al.*[10]

was unrelated to differences in pathological factors and was largely caused by:

- The high in-hospital mortality
- A morbidity that was carried on into the period after discharge from hospital

A policy of primary resection in obstruction had a similar in-hospital mortality to a policy of performing a conventional staged resection (primary resection, 19 per cent mortality; staged resection, 22 per cent mortality), even when considering left-sided cancers (left colon – primary resection, mortality 22 per cent; staged resection, mortality 20 per cent); advantages to a primary resection were a lower wound infection rate and a shorter hospital stay.[10]

It was also noted that obstructed tumours had a 21 per cent chance of subsequent local recurrence, compared with 13 per cent when obstruction was absent ($p<0.01$). This difference was not thought to be a reflection of advanced stage:

- Because even when stage was taken into account obstruction remained an independently significant prognostic variable
- Because a separate analysis of obstructing tumours did not show that they were on the whole either more advanced or more aggressive than non-obstructing tumours

It is always possible that high intraluminal pressure in obstruction forces cancer cells into lymphatic channels. During dissection these cells may leak from cut lymphatics and innoculate the wound. If clinically relevant, then it might be useful to dissect with a laser, which is known to seal lymphatics – unlike when either conventional dissection or electrocautery are used.[31] However, it seems much more likely that in the presence of obstruction a less than adequate cancer operation is sometimes undertaken. This may be in part because operations involving obstruction can be quite difficult (although preliminary decompression of the colon usually obviates such difficulty[32]), but it is more likely in the UK as less experienced surgeons perform the majority of the emergency surgery. For example, in the Large Bowel Cancer Project 43 per cent of emergency colonic resections were performed by Consultants compared with 75 per cent of elective colonic resections.[10]

The use of cytocidal agents in the peritoneal cavity when dealing with a perforated case seems sensible, although the effect of such agents on adhesion formation remains unknown. Mercuric perchloride should never be used in the peritoneal cavity as it may be absorbed and lead to renal failure. This leaves the choice between povidone-iodine, chlorhexidine/cetrimide, and low molecular weight dextran.

The extent of surgery

It is clear that a large proportion of local recurrences after rectal cancer surgery could be avoided by more extensive local surgery, and particularly by assiduous attention to the removal of the whole of the mesorectum[34] and an understanding of the geometry of the lateral ligaments. In part an improvement in results can be expected from training senior surgical registrars in the nuances of deep pelvic dissection. But there are claims that more extensive pelvic lymphadenectomy results not only in lower local recurrence rates but also in longer survival – although there is a trade off functionally because of inevitable sexual and urinary difficulties. Two studies are important in this regard: the first being from Warren Enker at the Memorial Sloan Kettering in New York;[35] and the second from Japan.[36] In the latter study, 232 patients are reported who underwent extended lymphadenectomy to remove the internal iliac and obturator lymph nodes at the time of rectal cancer surgery. There was only one postoperative death and the disease-free 5-year survival was 69 per cent overall. The lateral pelvic sidewall lymph nodes were involved by cancer in 18 per cent of cases overall, and in 36 per cent of Dukes' C cases, and yet the 5-year survival in patients with lateral nodal involvement was a very respectable 49 per cent. Against such extended lymphadenectomy are the following: 50 per cent of cases did not have nodal involvement so the extensive surgery was in retrospect unnecessary; the average operating time was 5.5 hours; average blood loss was 1.6 litres; and sexual and urinary disturbances were universal – although permanent catheterization was only required in 10 per cent.

The other issue to be addressed is that of high or low ligation of the inferior mesenteric artery. When performing an anterior resection the inferior mesenteric artery may either be ligated flush with the aorta or instead the left colic artery may be preserved. The background to this controversy is that a few patients with apical lymph node involvement will be cured of their cancer. For example, in the Large Bowel Cancer Project the age-adjusted 5-year survival for patients with apical lymph node involvement was 22 per cent.[37] The extent of surgery in these cases must have been sufficient either to remove the last residue of cancer or, alternatively, to tilt the immunological balance in the patients favour. Presumably there are times when the extra amount of vascular and lymphatic clearance achieved by high ligation might be sufficient to cure a patient who would not have been cured by low ligation. To date, however, it has not been possible to demonstrate any survival advantage to high ligation.[38] This may be because such cases also have had pelvic sidewall lymph node involvement that has been left untreated; the evidence presented from Japan in the previous paragraph lends some support to this supposition.

The surgeon who performs the operation

There is wide variation in the results achieved by different surgeons. Personal series, such as the one by Bill Heald, emphasize the sorts of results that can be achieved (n = 115, local recurrence in three[34]). Of course, such personal series are always open to the charge that they have concentrated on selected groups of patients. This is because even after curative abdominoperineal excision, where the extent of mesenteric and tissue clearance is certainly no less, experienced surgeons have been met by subsequent local recurrence in about 10 per cent of cases.[22] Furthermore, the finding from Japan that as many as 36 per cent of Dukes' C cases have pelvic sidewall lymph node involvement by cancer[36] means that the inevitable local recurrence rate is likely to be somewhat higher than 2–3 per cent. Nonetheless, in support of Heald's contention that technical skill is one of the most important factors in determining outcome, the Large Bowel Cancer Project was able to show that the surgeon who performed the operation is an independently significant predictor of later local recurrence, much in the same way as is the Dukes' stage.[16] 20 Consultant Surgeons, who between them had submitted between 30 and 105 curative operations into the study, were found to have local recurrence rates that ranged from less than 5 per cent to greater than 20 per cent. When surgeons with high local recurrence rates operated in favourable circumstances they still had poorer results than might have been expected on pathological grounds alone.

Medial to lateral dissection

Some colorectal cancers show venous vascular invasion. During mobilization, tumour emboli might be dislodged and give rise to subsequent liver metastases, a sequence of events that could be prevented by ligation of the vascular pedicle before handling the tumour. Turnbull[39] adopted such a technique and compared the survival of his patients with those of five other staff surgeons at the Cleveland Clinic who continued to perform a conventional dissection, which was characterized by mobilization of the tumour followed by vascular ligation. He found an improved survival, particularly for Dukes' C cases (58 per cent compared with 28 per cent), but such a difference might also have been because the extent of mesenteric clearance that he employed was much wider than that of his colleagues. Now the issue seems to have been resolved by a randomized controlled clinical trial from the Netherlands.[40] 236 patients showed no difference in 5-year survival between the two techniques of early or late vascular ligation ('no-touch', 60 per cent; conventional, 56 per cent). Nevertheless, it would not be unreasonable to expect early hepatic recurrence to be caused by the presence of occult hepatic metastases at the time of the original operation, and later ones to be caused by tumour emboli at the time of surgery. The study from the Netherlands has a fairly short follow-up that may itself explain the current lack of any survival advantage to early vascular ligation.

Blood transfusion

There are many reasons to avoid unnecessary blood transfusion. These include a fear of transmissible agents within the blood (e.g. AIDS, hepatitis, syphilis, etc.), fear of transfusion reactions, and a worry that through an effect on the immune system both wound infection rates and cancer recurrence rates might be increased. But blood transfusion reactions of any severity are very unusual, and blood is screened to exclude a number of potentially dangerous organisms (of course, this does not satisfy the Jeremiah instinct that is more concerned about the 'as yet undiscovered' adverse substances in blood, as was the case with the AIDS virus until the 1980s).

Insofar as an immune effect is concerned, it is now common knowledge that renal allograft recipients who have had blood transfusions are less likely to reject their kidneys. The problem for deciding the question for cancer recurrence is that the cases most likely to have a blood transfusion (e.g. low rectal cancers and bulky cancers) are also the cases that are most likely to do badly anyway. Quite a number of retrospective analyses have now examined this topic; some have found that blood transfusion has no effect on ultimate survival, others have found that transfused patients do worse.[42] The issue has recently been reviewed by Tartter,[43] who concludes that on the basis of the current studies no recommendations concerning the use of blood and colorectal cancer recurrence are justified.

It is very unlikely that this question will ever be resolved to everyone's satisfaction. A randomized study is fraught with practical and ethical difficulties: someone who needs blood must be given it; and it seems unreasonable that someone who does not need blood should be exposed to the other attendant risks of blood transfusion through a process of randomization. The issue is likely to become redundant anyway as in the USA. Surgeons there have experienced concern over the source and availability of blood for transfusion for a long time, and this has driven them to explore other ways of avoiding the **unnecessary** need for blood transfusion. The two most relevant methods are:

- In specialties where blood loss is largely unavoidable, for example, vascular surgery, blood cell recovery and autotransfusion
- Where blood loss can be either avoided or severely restricted, as is the case with most colorectal operations, the use of coagulating diathermy rather than knife and scissors for the dissection

It is highly likely that both of these approaches will become equally popular on this side of the Atlantic.

Conclusion

Despite great strides nothing yet exists in modern science and therapeutics that can have as great an impact on patients' well being after colorectal cancer surgery as: the general condition of the patient at presentation; careful preoperative preparation; and the skill of the surgeon in the operating theatre.

References

1 Slaney G. In Irvine W (Ed), *Modern trends in surgery* (Vol 3). Sevenoaks: Butterworths, 1971, pp. 69–89.
2 Finlay I G, McArdle C S. Occult hepatic metastases in colorectal carcinoma. *Br J Surg*. 1986; **73**: 732–5.
3 Fielding L P, Phillips R K S, Hittinger R. Factors influencing mortality after curative resection for large bowel cancer in elderly patients. *Lancet*. 1989; **1**: 595–7.
4 Lunn J N, Devlin B H. Lessons from the Confidential Enquiry into Perioperative Deaths in three NHS regions. *Lancet*. 1987; **2**: 1384–6.
5 Lancet editorial. Accounting for perioperative deaths. *Lancet*. 1987; **2**: 1369–71.
6 Aldridge M C, Phillips R K S, Hittinger R, *et al*.. Influence of tumour site on

presentation, management and subsequent outcome in large bowel cancer. *Br J Surg*. 1986; **73**: 663–70.
7 Lewis A A M, Khoury G A. Resection for colorectal cancer in the very old: are the risks too high? *Br Med J*. 1988; **296**: 459–61.
8 Morel Ph, Egeli R A, Wachtl S, Rohner A. Results of operative treatment of gastrointestinal tract tumours in patients over 80 years of age. *Arch Surg*. 1989; **124**: 662–4.
9 Lindmark G, Pahlman L, Enblad P, Glimelius B. Surgery for colorectal cancer in elderly patients. *Acta Chir Scand*. 1988; **154**: 659–63.
10 Phillips R K S, Hittinger R, Fry J S, Fielding L P. Malignant large bowel obstruction. *Br J Surg*. 1985; **72**: 296–302.
11 Fielding L P, Stewart-Brown S, Blesovsky L, Kearney G. Anastomotic integrity after operations for large bowel cancer: a multicentre study. *Br Med J*. 1980; **281**: 411–4.
12 Wangensteen O H. Cancer of the colon and rectum – with special reference to: 1) earlier recognition of alimentary tract malignancy; 2) secondary delayed re-entry of the abdomen in patients exhibiting lymph node involvement; 3) subtotal primary excision of the colon; 4) operation in obstruction. *Wis Med J*. 1949; **48**: 591–7.
13 Gilbertsen V A. Adenocarcinoma of the rectum: incidence and locations of recurrent tumor following present-day operations performed for cure. *Ann Surg*. 1960; **151**: 340–8.
14 Gunderson L L, Sosin H. Areas of failure found at re-operation following 'curative' surgery for adenocarcinoma of the rectum. *Cancer*. 1974; **34**: 1278–92.
15 Pahlman L, Glimelius B. Local recurrences after surgical treatment for rectal carcinoma. *Acta Chir Scand. 1984;* **150**: 331–5.
16 Phillips R K S, Hittinger R, Blesovsky L. Local recurrence following 'curative' surgery for large bowel cancer. The overall picture. *Br J Surg*. 1984; **71**: 12–16.
17 Gilchrist R K, David V C. A consideration of pathological factors influencing 5-year survival in radical resection of the large bowel and rectum for carcinoma. *Ann Surg*. 1947; **126**: 421–38.
18 Stearns M W, Binkley G E. The influence of location on prognosis in operable rectal cancer. *Surg Gynaecol Obstet*. 1953; **56**: 368–70.
19 Garlock J H, Ginsburg L. An appraisal of the operation of anterior resection for carcinoma of the rectum and rectosigmoid. *Surg Gynecol obstet*. 1950; **90**: 525–34.
20 Judd E S, Bellegie N J. Carcinoma of rectosigmoid and upper part of rectum (recurrence following low anterior resection). *Arch Surg*. 1952; **64**: 697–706.
21 Deddish M R, Stearns M W. Anterior resection for carcinoma of the rectum and rectosigmoid area. *Ann Surg*. 1961; **154**: 961–6.
22 Morson B C, Vaughan E G, Bussey H J R. Pelvic recurrence after excision of rectum for carcinoma. *Br Med J*. 1963; **2**: 13–19.
23 Goligher J C. *Surgery of the Anus, Rectum and Colon* (4th edn). London: Baillière Tindall, 1980, p 431.
24 Theile D E, Cohen J R, Evans E B, *et al.*. Pelvic recurrence after curative resection for carcinoma of the rectum. *Aust NZ J Surg*. 1982; **52**: 391–4.
25 Rankin F W, Olson P F. The hopeful prognosis in cases of carcinoma of the colon. *Surg Gynecol Obstet*. 1933; **56**: 366–74.
26 Irvin T T, Greaney M G. The treatment of colonic cancer presenting with intestinal obstruction. *Br J Surg*. 1977; **64**: 741–4.
27 Shepherd J M, Jones J S. Adenocarcinoma of the large bowel. *Br J Cancer*. 1971; **25**: 680–90.
28 Olson R M, Perencevich M P, Malcolm A W. Patterns of recurrence following

curative resection of adenocarcinoma of the colon and rectum. *Cancer*. 1980; **45**: 2969–74.

29 Griffiths J D. Surgical anatomy of the blood supply of the distal colon. *Ann R Coll Surg Engl*. 1956; **19**: 241–56.

30 Ohman U. Prognosis in patients with obstructing colorectal carcinoma. *Am J Surg*. 1983; **143**: 742–7.

31 Zimmerman I, Stern J, Frank F, *et al.*. Interception of lymphatic drainage by Nd YAG laser irradiation in rat urinary bladder. *Lasers Surg Med*. 1984; **4**: 167–72.

32 Dudley H A F, Phillips R K S. Intraoperative techniques in large bowel obstruction: methods of management with bowel resection. In Fielding L P, Welch J (Eds), *Intestinal Obstruction. Clinical Surgery International* (Vol. 13). Edinburgh, London, Melbourne, New York: Churchill Livingstone, 1987.

33 Umpleby H C, Williamson R C N. The efficacy of agents employed to prevent anastomotic recurrence in colorectal carcinoma. *Ann R Coll Surg*. 1984; **66**: 192–4.

34 Heald R J, Ryall R D H. Recurrence and survival after total mesorectal excision for rectal cancer. *Lancet*. 1986; **2**: 1479–82.

35 Enker W E, Pilipshen S J, Heilwell M L, *et al.*. En bloc pelvic lymphadenectomy and sphincter preservation in the surgical management of rectal cancer. *Ann Surg*. 1986; **293**: 426–33.

36 Moriya Y, Hojo K, Sawada T, Doyama Y. Significance of lateral node dissection for advanced rectal carcinoma at or below the peritoneal reflection. *Dis Colon Rectum*. 1989; **32**: 307–15.

37 Phillips R K S, Hittinger R, Blesovsky L. Large bowel cancer: surgical pathology and its relationship to survival. *Br J Surg*. 1984; **71**: 604–10.

38 Pezim M F, Nicholls R J. Survival after high and low ligation of the inferior mesenteric artery during curative surgery for rectal cancer. *Ann Surg*. 1984; **200**: 729–33.

39 Turnbull R B, Kyle K, Watson F R, Spratt J. Cancer of the colon: the influence of the no-touch technique on survival rates. *Ann Surg*. 1967; **166**: 420–7.

40 Wiggers T, Jeekel J, Arends J W, *et al.*. No-touch isolation technique in colon cancer: a controlled prospective trial. *Br J Surg*. 1988; **75**: 409–15.

41 Opelz G, Sengar D P S, Mickey M R, Terasaki P I. Effect of blood transfusions on subsequent kidney transplants. *Trans Proc*. 1973; **5**: 253–9.

42 Parrott N R, Lennard T W J, Taylor R M R, *et al.*. Effect of perioperative blood transfusion on recurrence of colorectal cancer. *Br J Surg*. 1986; **73**: 970–3.

43 Tartter P I. Perioperative blood transfusion and colorectal cancer recurrence: a review. *J Surg Oncol*. 1988; **39**: 197–200.

7

Treatment of acute diverticulitis of the colon

Sven Goldman and Thomas Ihre

Introduction

The latin word 'diverticulum' was used in Rome to describe 'a small place by the side of the road with a bad reputation'. Although the original description has been somewhat modified, its original meaning implying an association with complications still holds true. A diverticulum of the colon is an acquired lesion created by a herniation of the mucosa and submucosa through a defect in the circular muscular layer where the vessels penetrate to supply the mucosa. The diverticula are referred to as 'false' as they have no supporting muscle layers in the diverticulum. They are usually located between the mesenteric and antimesenteric taenia. If more than one diverticulum is present in the absence of inflammation, the condition is often referred to as 'diverticulosis', and when inflammation has occurred the term 'diverticulitis' is used. The herniation is claimed to be caused by high pressure inside the colon produced by hypertrophy of the circular muscle layer.[1] It is widely accepted that the main cause of this condition is a deficiency of fibre in the normal diet. This assumption is supported by the fact that in Africa and Asia, where the content of fibre in the diet is much higher, the incidence of diverticulosis is low.[2] Furthermore, immigrants from these areas, once they have adopted the 'Western' type of diet, develop diverticulosis in almost the same frequency as Westerners.

Diverticulosis of the colon is a common condition in the western hemisphere. The incidence of diverticulosis increases with age. It is uncommon before the age of 40, but evidence has been presented indicating that up to 70 per cent of a Western population will have some degree of diverticulosis by the age of 80.[3] Males and females are equally affected. The most common site of diverticulosis is the sigmoid colon. Right-sided diverticular disease ranges from 0.1–2.5 per cent of all cases and is more common in Asian countries, especially Japan. These patients are also 10–15 years younger than patients with left-sided disease.[4]

The symptoms of diverticulosis are related to high pressure inside the intestine – abdominal pain, alteration of bowel habit and sometimes even nausea and vomiting. These symptoms are similar to those seen in the irritable bowel syndrome, which has been looked upon, by some, as a precursor of diverticulosis. In both these conditions laboratory tests give normal results. Occasionally a spastic segment of the sigmoid colon is palpable. The diagnosis of diverticulosis is based on the findings of a barium enema examination or a colonoscopy.

Complications of diverticulosis are bleeding, which is often slight but sometimes massive, and diverticulitis, which may then lead on to secondary complications such as abscess, perforation, fistula or stenosis.

In patients with ongoing massive rectal bleeding, having excluded upper gastrointestinal causes and bleeding haemmorrhoids, the diagnostic/management choices are: selective angiography and colonoscopy. Selective angiography has the advantages that it may lead to precise preoperative localization of the site of bleeding and can rule out other sources of bleeding such as angiodysplasia[6] or ischaemic colitis[7] that may well coexist with diverticulosis, especially in elderly patients. By injection of pitressin through the angiography catheter, control of the bleeding can be achieved in up to half of the patients. It is estimated that the bleeding must exceed 0.5 ml/minute if leakage of contrast into the bowel lumen is to be demonstrable.[8] With radiolabelled red blood cell scan it is possible to pick up a lower rate of bleeding (0.25 ml/minute), but with this technique it is difficult to visualize areas with angiodysplasia. Colonoscopy has been used in some cases to diagnose the site of bleeding. However, this technique is difficult to perform as it prerequisites sufficient bowel preparation, which is difficult to achieve as the bleeding in most cases is continuous. Colonoscopy during these circumstances has been looked upon as risky by some surgeons. If the site of bleeding has not been possible to diagnose in a patient with diverticulosis and if the patient continues to bleed massively, urgent laparotomy is indicated. This will then include a blind resection of that part of the colon that is affected by the diverticulosis. If a primary anastomosis is performed, many surgeons will make a temporary transverse colostomy so that if the patient continues to bleed, the bleeding source can be diagnosed to the right or left side of the transversostomy. An additional life-saving laparotomy will then include resection of the part of colon that is bleeding.

In patients with diverticulitis the inflammation may cause severe pain and localized peritonitis, as well as strictures. The strictures may require endoscopy to rule out cancer. In some patients the inflammation will lead to perforation of the diverticulum, with generalized peritonitis. About 5–10 per cent of all patients with diverticulitis will require emergency surgery.[9] The main topic of this chapter is the surgical management of patients with acute diverticulitis.

Acute diverticulitis

The pathogenesis of an acute attack of diverticulitis is not known. The question whether this inflammation occurs primarily within a diverticulum that perforates into surrounding tissues or whether it develops from mechanical effects on one or more diverticulae, leading to mucosal damage and even perforation with secondary inflammation, remains controversial. The latter explanation would seem more likely to us, as diverticulitis begins mostly with an inflammation around one single diverticulum. If this diverticulum perforates, other diverticulae may become involved secondarily through the spread of inflammation within the pericolic fat, and an abscess may form alongside the bowel wall, giving rise to localized peritonitis and impairment of the passage of bowel contents. Intestinal obstruction may also occur and is generally believed to be caused by oedema surrounding an inflamed diverticulum. In most patients such obstruction is transient. Localized peritonitis develops when, in the event of a perforation, the leakage is limited by the omentum or by previous adhesions to nearby intestine or organs, and may result in the formation of a fistula to the adherent organ. In patients without such adhesions the perforation results in leakage of intestinal contents into the peritoneal cavity and generalized peritonitis. Most of these patients will be in a state of septic shock on admission and will require adequate treatment with intravenous fluid and antibiotics before surgery.

Symptoms

The subjective symptoms of acute diverticulitis include severe pain in the lower left and/or right quadrant accompanied by acute tenderness and guarding in the same region. A mass is sometimes palpable, and fever and cessation of passage of gas and faeces are also common symptoms. When perforation of the diverticulum has resulted in a fistula, the symptoms depend on which organs are involved. Of the faecal fistulas that may be produced in this way, the colovesical fistula is the most common (Fig. 7.1). This has been reported to occur in 2–4 per cent of male patients with diverticulitis, but to be rare in female patients, in whom colovaginal or colo-uterine fistulas may be formed, except in women who have undergone a hysterectomy.[10,11] The urinary symptoms, which include passage of gas with the urine, may often be the first sign of the underlying diverticular disease. Passage of faecal material through the urethra is rare and the urine may well be sterile despite the presence of bacteria in the colon. The fistula may be difficult to visualize radiologically, irrespective of whether a barium enema or a retrograde uretherocystography is performed. However, these examinations need to be performed as the fistula may arise from other diseases, such as carcinoma of the colon, rectum or the bladder, or from terminal ileal Crohn's disease. Often the patient's symptoms and the observation of diverticulitis during barium examination are the only positive findings preceding an operation. The majority of male patients who develop a colovesical fistula have had few or no symptoms of their diverticular disease before the perforation. This condition could be managed electively with a one-stage resection and anastomosis in the majority of cases. If no opening into the bladder is visible during laparotomy urethral catheter drainage for 10–14 days is sufficient. Visible perforations into the bladder wall are sutured with two layers of absorbable sutures. A pedicle of omentum should be interposed between the colonic anastomosis and the bladder if possible.

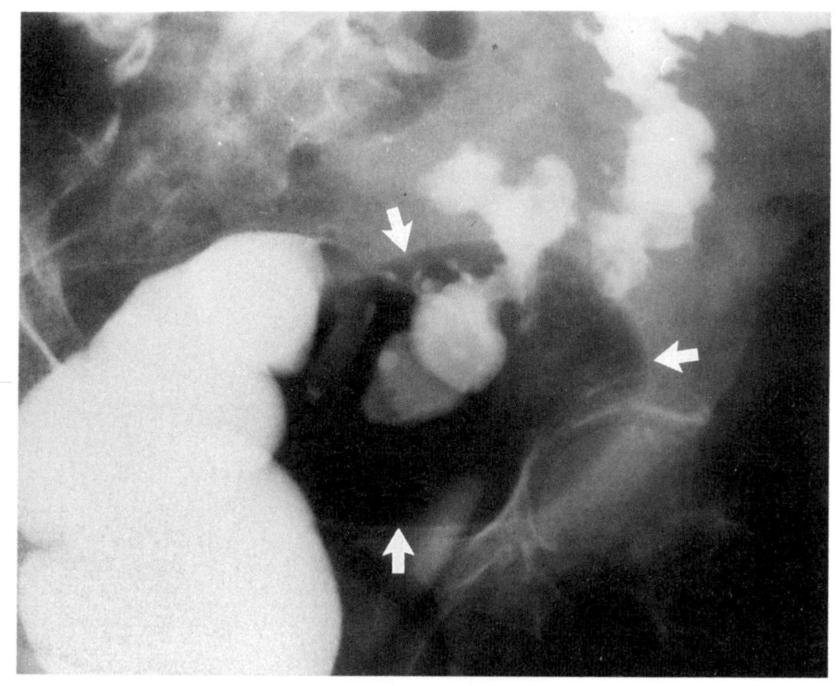

Fig. 7.1 Barium enema showing diverticulitis of the sigmoid colon and gas in the bladder (arrows) caused by a colovesical fistula.

Acute complete obstruction of the sigmoid colon by diverticulitis is rare, but some degree of chronic obstruction is common. Symptoms of acute obstruction of the colon are more apt to be produced by secondary inflammation of the small intestine or as a result of formation of an inflammatory diverticular mass with involvement of parts of the small intestine. Primary resection and proximal colostomy according to Hartmann operation is usually the safest procedure. Other options include transverse colostomy and caecostomy.

Diagnosis

In a survey presented by Dawson *et al.*[12] of a series of 93 patients operated on for perforated diverticulitis, only one-third were correctly diagnosed preoperatively. The most common false preoperative diagnosis was appendicitis, which was suspected in one-third of the patients. The diagnosis is usually established at a laparotomy necessitated by septic complications and peritonitis. Apart from the clinical history and examination and the conclusions that can be drawn from these, plain abdominal radiography will sometimes be helpful and show free gas, thus indicating perforation into the peritoneal cavity. An enema with a water-soluble contrast medium may in some cases reveal the perforation through the leakage of contrast (Fig. 7.2). Some authors have advocated the use of computed tomography

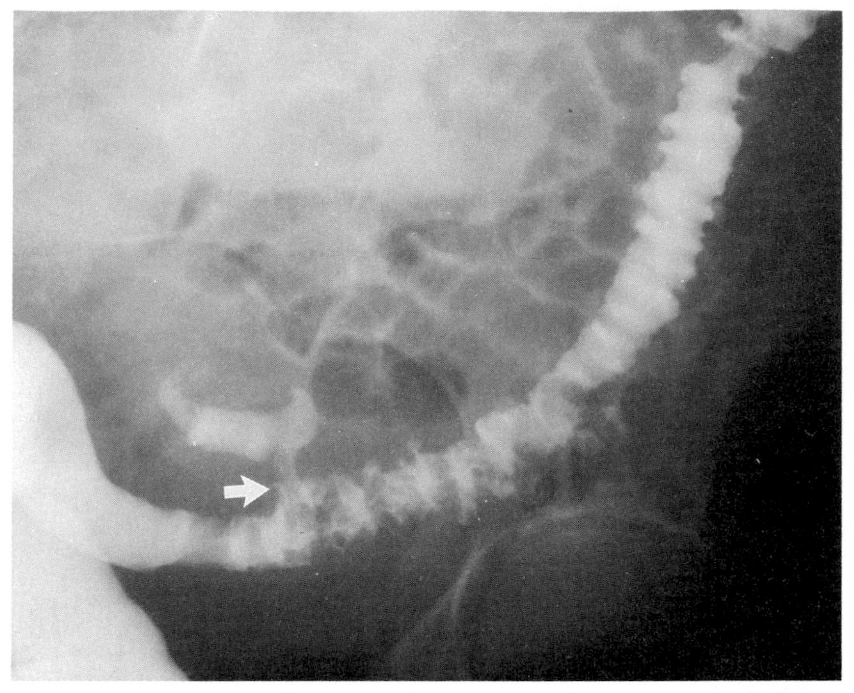

Fig. 7.2 Contrast leakage through a perforation of a diverticulum of the sigmoid colon (arrow).

(CT) as the safest method of diagnosing perforated diverticulitis.[13,14] The typical findings with this technique include localized thickening of the intestinal wall and an increasing density of the pericolic fat. This method could then be used in patients with suspected diverticular complications in whom the diagnosis is uncertain. Other authors have suggested that sonography is a valuable imaging technique in a patient with signs and symptoms of acute complicated diverticulitis. However, sonographic descriptions of patients with diverticular disease are limited to a few reports: one of caecal diverticulitis[15] and two descriptive articles of sigmoid diverticulitis.[16,17]

Treatment

It is generally accepted that about one-third of all patients with diverticular disease will develop inflammatory complications and about one-third of these will have persistent symptoms.[9] In a study presented by Boles and Jordan, who followed 294 patients with diverticular disease for a period of 10–30 years, 38 per cent of the patients with diverticulitis developed inflammatory complications.[18] These symptoms include recurrent attacks of abdominal pain with signs of bacteraemia, fistulas, stenosis of the sigmoid colon, perforation and rarely bleeding. In patients without perforation, conservative treatment is usually sufficient. In a Swedish study of

392 patients with acute sigmoid diverticulitis, all 295 patients without perforation were successfully treated without operation.[19] Hospitalization is advocated, as there may be complications demanding immediate surgical intervention. If bacteraemia develops, systemic antibiotic therapy covering both aerobic and anaerobic microorganisms is indicated. Even if the diverticulitis episode is uncomplicated the risk of recurrence is great and if medical treatment fails to give permanent improvement in these patients surgery is indicated. The estimated risk for later recurrence has been reported to be 40–45 per cent.[18,20,21] The calculated risk of having recurrent disease within the first year after the initial attack has been estimated as 10 per cent.[19] Surgery is also indicated in patients in whom the barium enema and colonoscopy have failed to distinguish between diverticulitis and cancer. The presence of diverticula in the bowel above or below a stricture favours a diagnosis of diverticulosis, but does not rule out the presence of cancer. In some series of patients undergoing an emergency operation for acute diverticulitis, the coincidence rate of colorectal cancer in the resected specimen has been found to be as high as 25 per cent.[22] In our own study of 107 patients operated upon for acute sigmoid diverticulitis, 6 per cent had colorectal malignancy.[23]

According to Rodkey and Welch,[24] definite indications for elective surgery are:

- Recurrent attacks of local inflammation
- Persistent tender mass
- Stenosis on roentgenographic examination
- Clinical or roentgenographic signs too equivocal to rule out carcinoma
- Clinical or roentgenographic signs of fistula

They also include other symptoms associated with complications of diverticulosis as relative indications, namely dysuria, functional colonic disturbances and abdominal discomfort, and rapid progression of symptoms from the time of onset. An additional indication is a relatively young age of the patient, below 50 years.

Elective surgical procedures in diverticulitis

Depending on the severity of the diverticular disease, two different procedures can be used; primary resection and anastomosis with or without a covering colostomy on the transverse colon, and the Hartmann operation. The latter includes a colostomy and a closed rectal stump. Elective resection should be performed 9–12 weeks after an acute episode. The time interval permits resolution of the inflammatory reaction. The preparation of patients for elective colon resection should include both mechanical and antibiotic preparation. If, at an operation with a primary resection and anastomosis, an abscess is present the risk of anastomotic leakage increases and a diversion of faeces by a transverse colostomy is advisable. In patients with no abscess formation and in whom the indication for surgery is recurrent attacks of diverticulitis or uncertainty regarding malignancy, a primary anastomosis without a temporary colostomy may well be performed with little or no risk to the patient. The mortality with primary resection and anastomosis with these indications has been reported to be as low as 1.4–2.1 per cent.[25]

Surgical treatment in acute complicated diverticulitis
During the past decades several surgical procedures have been used in patients with peritonitis, sometimes including faecal contamination, resulting from perforated diverticulitis. These have included: simple drainage with or without a covering defunctioning transverse colostomy; suturing the perforation or covering it with omentum with or without a covering proximal colostomy; a primary resection with anastomosis with or without a temporary transverse colostomy or a primary resection without anastomosis and proximal colostomy (Hartmann type). At one time exteriorization of the perforated segment was performed. Some of the above procedures are combinations performed in three stages, beginning with a proximal colostomy and followed later by a resection of the diseased bowel, and at the third stage closure of the stoma. Experience showed that this regimen was associated both with a high morbidity and mortality and with long periods of hospitalization.[26,27] Moreover, the frequency of closure of the stoma was low.[22,27] The high mortality rate in these patients was mainly due to sepsis from the peritonitis caused by the diverticulitis. Krukowski and Matheson[22] reported on a large series of 1292 patients with complicated diverticular disease. The mortality rate among patients undergoing delayed resection was 26 per cent, but among those who underwent resection without anastomosis it was 12 per cent. Primary resection with anastomosis with or without colostomy carried the lowest mortality of only 8 per cent. Similar results have been presented by Greif *et al.*[28] in a retrospective review of 1353 cases. Killingback,[29] however, reported an anastomotic leakage rate of 30 per cent in patients treated by resection and primary anastomosis.

The Hartmann operation with an end-colostomy and distal oversew of the intraperitoneal rectum has for many years usually been the method of choice for these critically-ill patients. This operation has the advantage of removing the diseased bowel without a risk of anastomotic leakage, which seems to be correlated with a high mortality rate.[30] It is now generally accepted that in patients with peritonitis it is important that the affected perforated segment is removed. It is often difficult to discover the site of perforation, which may well be covered; often the sigmoid colon is enlarged and swollen and the resection has to be done with margins to ensure that the perforated bowel is removed.

Technical aspects of the Hartmann operation

The following approach has been used in patients with acute complicated diverticulitis at the Department of Surgery of Södersjukhuset, Stockholm, Sweden during the past 10 years.

No bowel irrigation is carried out preoperatively. Nor is preoperative retrograde ureteral catheterization performed routinely. All patients receive pre- and peroperative infusion of broadspectrum antibiotics covering both aerobic and anaerobic microorganisms. This regimen is continued for 5–7 days postoperatively.

A low midline abdominal incision is made. After division and ligation of the superior haemorrhoidal artery and branches of the inferior mesenteric arteries, the distal resection line is usually at the level of the sacral promontory. The pelvic peritoneum is not opened. The rectal stump is closed either by a single layer of interrupted absorbable suture or by a linear stapler. After division and resection of the sigmoid segment, the proximal part of the colon is brought out through an opening in the left rectus muscle. The seromuscular coat of the colon is fixed to the parietal peritoneum and the stoma is fixed to the skin by mucocutaneous sutures. Low-dose heparin is routinely administered postoperatively for 7 days.

After an interval of approximately 3 months the excluded rectum and the proximal colon are investigated by sigmoidoscopy and barium enema, and if satisfactory, restoration of the colonic continuity is then performed. Preoperatively patients receive metronidazole and doxycycline orally the day before and the morning of surgery. A single bolus dose of metronidazole is also given intravenously after operation. Low-dose heparin is given prophylactically before surgery and for 7 days postoperatively. The colorectal anastomosis is either achieved with a single layer of interrupted sutures or with an EEA stapling device. The latter procedure has proved useful in cases where identification of the rectal stump is difficult or when retraction of the stump has occurred.

Operative results in acute diverticulitis

The outcome of surgery in these critically-ill patients depends on a number of factors, including the age of the patient, the type of pre-, per- and postoperative treatment and the degree of peritoneal contamination. To review historical data and compare them with today's results is almost impossible – the antibiotics that are available today are much more potent than those available a few years ago; and better understanding of fluid and electrolyte balance and more advanced anaesthetic techniques have also helped.

Another obstacle to comparison of the results of different operative procedures given in different reports is the lack of standardized classifications of disease severity. The major criticism of the Hartmann procedure, when compared with primary resection and anastomosis have been the reports on higher morbidity and mortality and the fact that more than half of the patients have been left with a permanent stoma.[31–33] The degree of contamination in the abdomen (e.g., whether free perforation has occurred or only an abscess is present) will influence the outcome of any operation.[34] The choice of operative method by the surgeon is likely to be influenced by the extent of contamination, which could then bias evaluation of a specific operative procedure.

In order to overcome this, in a retrospective study in Stockholm,[23] a modification of the classification of Hinchey *et al.*[35] in acute diverticulitis was used. On the basis of the findings at operation the patients were assigned to four stages:

Stage I: Non-perforated inflammatory mass in the sigmoid colon
Stage II: Mesenteric or pericolic abscess with no visible perforation
Stage III: Generalized purulent peritonitis with identified free perforation of a pericolic abscess
Stage IV: Generalized faecal peritonitis with identified free perforation of the sigmoid colon

The study comprised all the 107 patients (61 women and 46 men, median age 65 years), who were operated upon for acute sigmoid diverticulitis over a 10-year period (1979–1988). Coexisting disease is shown in Table 7.1. According to the medical records 50 patients belonged to Stages I and II, and 48 to Stages III and IV and in nine patients stageing could not be assessed because of inadequate recording (Table 7.2). In 78 per cent of the patients localized or generalized peritonitis was the indication for operation and in 22 per cent intestinal obstruction. It was notable that 92 (86 per cent) of the 107 patients had no previous history of diverticulitis. None of the 22 patients in Stage IV had had previous attacks of diverticulitis.

The Hartmann procedure was used in 36 patients with Stages I and II, in 43 of the patients with Stages III and IV and in one unclassified patient. Primary resection with anastomosis was performed in 11 patients in Stages I and II (one had a covering colostomy), in two patients in Stage IV (one had a covering colostomy) and in two unclassified patients. Transverse colostomy was the only operative procedure in three patients in Stage II, in two patients in Stage III and in six unclassified patients. Drainage alone was used in one patient with Stage III disease. Altogether there were ten deaths, giving an overall mortality of 9 per cent (Table 7.3). Seven of these deaths occurred in Stage IV patients – all had been operated on by the Hartmann procedure. Two deaths occurred among the nine unclassifiable patients. Both these patients had been treated with a transverse colostomy alone. One patient in Stage I, who had undergone a primary resection and anastomosis, died of cerebral thrombosis 3 weeks after the operation. The main causes of death were cardiac failure and septicaemia, which were combined in two patients. Most deaths occurred in Stage IV patients, i.e. those with faecal peritonitis, where the mortality rate was 32 per cent. The majority were older than 75 years (see Table 7.4).

The morbidity after operation for acute diverticulitis was high, 54 per cent in the group with a Hartmann operation, 27 per cent in the group with primary resection, and 45 per cent in the remaining patients (see Table 7.5). The most frequent postoperative complications were cardiopulmonary, stoma necrosis, anastomotic leakage and wound infection (Table 7.6).

Of the 73 surviving patients operated on by the Hartmann procedure, reconstruction was attempted in 57 (78 per cent). Five had to undergo subsequent reoperation because of anastomotic leakage. There were no postoperative deaths in this group of patients. Of the 16 patients in whom no reconstruction was performed, five died of other diseases, four refused an additional operation and in five patients the general condition was

Table 7.1 Medical history in relation to stage of disease

Coexisting disease	All patients		Stage I		Stage II	
	$n = 107$		$n = 13$		$n = 37$	
	Number	(%)	Number	(%)	Number	(%)
Cardiovascular disease	44	(41)	6	(46)	17	(46)
Malignant disease	7	(6)	1	(8)	–	
Pulmonary disease	3	(3)	–		1	(3)
Diabetes	2	(2)	1	(8)	1	(3)
Steroid therapy	2	(2)	–		1	(3)

Coexisting disease	Stage III		Stage IV		Unstaged	
	$n = 26$		$n = 22$		$n = 9$	
	Number	(%)	Number	(%)	Number	(%)
Cardiovascular disease	5	(19)	11	(50)	5	(55)
Malignant disease	–		6	(27)	–	
Pulmonary disease	1	(4)	–		1	(11)
Diabetes	–		–		–	
Steroid therapy	–		1	(3)	–	

judged too poor to permit reconstruction. Information was lacking for two patients.

From the Stockholm survey the following conclusions can be drawn: The Hartmann operation is a safe procedure with a low mortality rate in patients with Stage I, II or III disease. Faecal peritonitis (Stage IV disease) was associated with a high mortality rate (32 per cent) in spite of modern antibiotics. Most patients who died were elderly and had a history of cardiovascular disease. None of the patients with faecal peritonitis had had any previous attacks of diverticulitis. In the Hartmann group of patients the postoperative complication rate was high (54 per cent), probably because most of these patients had Stage III or IV disease (Table 7.5). Restoration of colonic continuity was possible in approximately 80 per cent of the patients. The overall mortality rate was low, confirming the advisability of resecting the inflamed sigmoid colon. The results do not contradict the view that primary resection with anastomosis may be used in selected cases with localized abscess formation (Stage II), but the series is too small to allow firm conclusions.

Table 7.2 Patient characteristics in relation to stage of disease

Stage of disease	Number of cases	(% of cases)	Sex male/female	Age in years	(range)	Number aged ≥ 75 years	(%)
Stage I	13	(12)	4/9	63	(43–74)	–	
Stage II	37	(35)	13/24	63	(36–88)	9	(24)
Stage III	26	(24)	15/11	61	(36–78)	3	(11)
Stage IV	22	(21)	9/13	75	(46–89)	11*	(50)
Unstaged	9	(9)	4/5	75	(62–86)	5	(56)
Total	107		45/62	65 (median)	(36–89) (range)	28	(21)

*Stage IV versus Stage I–III: $p < 0.01$.

Table 7.3 Surgical procedure and results in relation to stage of disease

Stage of disease	Hartmann's procedure		Resection with anastomosis		Transverse colostomy		Drainage		Postoperative complications		Length of hospital stay		Deaths	
	Number	(%)	Number	(%)	Number	(%)	Number	(%)	Number	(%)	Days	(Range)	Number	(%)
Stage I	8	(62)	5	(38)	–		–		7	(54)	16	(10–74)	1	(8)
Stage II	28	(76)	6*	(16)	3	(8)	–		15	(40	20	(8–68)	–	
Stage III	23	(88)	–		2	(8)	1	(4)	11	(42)	19	(10–61)	–	
Stage IV	20	(91)	2*	(9)	–		–		14**	(64)	21	(8–83)	7	(32)
Unstaged	1	(11)	2	(22)	6	(67)	–		5	(56)	17	(11–47)	2	(22)
Total	80	(75)	15	(14)	11	(10)	1	(1)	52	(49)	20	(8–83)	10	(9)

*Thereof one patient with protective colostomy. **Stage IV versus Stage I–III, $p < 0.05$.

Table 7.4 Details of the postoperative deaths (within 30 days)

Sex	Age (years)	Coexisting morbidity	Stage of diverticular disease	Procedure	Factors contributing to death	Dead on postoperative day
Female	75	None	IV	Hartmann	Cardiac failure	0
Male	87	Cardiovascular disease	IV	Hartmann	Cardiac failure	1
Female	76	Cardiovascular disease	IV	Hartmann	Septicaemia, cardiac failure	2
Male	75	Polio	–	Transverse colostomy	Septicaemia, cardiac failure	2
Male	84	Cardiovascular disease	–	Transverse colostomy	Pulmonary embolus	5
Female	57	Lymphoma	IV	Hartmann	Pneumonia, respiratory failure	7
Male	89	Cardiovascular disease, diabetes	IV	Hartmann	–	7
Female	80	Bladder cancer	IV	Hartmann	Cardiac failure	8
Male	61	Brain tumour, steroid therapy	IV	Hartmann	Septicaemia	18
Female	72	Cardiovascular disease, diabetes	I	Resection and anastomosis	Cerebral thrombosis	21

Table 7.5 Results in relation to surgical procedure

			Length of operation		Postoperative complications		Length of hospital stay		Deaths	
Surgical procedure	Number	(%)	Minutes	(range)	Number	(%)	Days	(range)	Number	(%)
Hartmann's procedure	80	(75)	140	(75–300)	43	(54)	21	(8–83)	7	(9)
Resection and anastomosis	15*	(14)	120	(85–170)	4	(27)	18	(10–42)	1	(7)
Transverse colostomy	11	(10)	60	(30–165)	5	(45)	20	(11–47)	2	(18)
Drainage	1	(1)	120		–		14		–	
Total	107		130	(30–300)	52	(49)	20	(8–83)	10	(9)

*Thereof two patients with protective colostomy.

Table 7.6 Complications in relation to surgical procedure

Complication	Hartmann's procedure	Resection with anastomosis	Transverse colostomy	Patients reoperated
		Number (%)		Number
Cardiopulmonary	14 (17)	2 (13)	1 (9)	–
Complications of colostomy	9 (11)	–	2 (18)	2
Wound infection	7 (9)	–	1 (9)	–
Abscess	4 (5)	–	1 (9)	3
Septicaemia	5 (6)	1 (7)	–	–
Postoperative ileus	5 (6)	–	–	2
Wound dishiscence	4 (5)	–	–	4
Thrombosis, embolia	2 (2)	–	1 (9)	–
Anastomotic leakage	–	2 (13)	–	2
Ischaemic bowel necrosis	2 (2)	–	–	2
Stump leakage	1 (1)	–	–	1
Bleeding	1 (1)	–	–	1
Ureteric lesion	1 (1)	–	–	–
Other	6 (7)	–	–	–
Complications per patient	1.3	1.0	1.2	

References

1 Ryan P. Changing concepts in diverticular disease. *Dis Colon Rectum*. 1983; **26**: 12–18.
2 Painter N S, Burkitt D P. Diverticular disease of the colon: A 20th century problem. *Clin Gastroenterol*. 1975; **4**: 3–22.
3 Hughes C E. Post-mortem survey of diverticular disease of the colon. *Gut*. 1969; **10**: 336–51.
4 Schuler J G, Baglay J. Diverticulitis of the caecum. *Surg Gynecol Obstet*. 1983; **156**: 743–8.
5 Hill G J, Taubman J O. Localized microscopic diverticulitis revealed by arteriography at the site of haemorrhage in diverticular disease of the colon. *Am J Dig Dis*. 1973; **18**: 808–12.
6 Whitehouse G H. Solitary and angiodysplastic lesions in the ileo-caecal region diagnosed by angiography. *Gut*. 1973; **14**: 977–82.
7 Marston A, Pheils M T, Thomas M L, Morson B C. Ischaemic colitis. *Gut*. 1966; **7**: 1–5.
8 Baum S, Nusbaum M. The control of gastrointestinal haemorrhage by selective mesenteric arterial infusion of vasopressin. *Radiology*. 1971; **98**: 497–505.
9 Chappuis C W, Cohn J, Jr. Acute colonic diverticulitis. *Surg Clin North Am*. 1988; **68**(2): 302.
10 Steele M, Deveney C, Burchell M. Diagnosis and management of colovesical fistulas. *Dis Colon Rectum*. 1979; **22**: 27–30.
11 McConnell D B, Sasaki T M, Velto R M. Experience with colovesical fistula. *Am J Surg*. 1980; **140**: 80–84.
12 Dawson J L, Hanon J, Roxburgh R A. Diverticulitis coli complicated by diffuse peritonitis. *Br J Surg*. 1965; **52**: 354–7.
13 Hulnick D H, Megibow A J, Balthazar E J, *et al.*. Computed tomography in the evaluation of diverticulitis. *Radiology*. 1984; **152**: 491–5.
14 Morris J, Stellato T A, Haaga J R, *et al.* The utility of computed tomography in colonic diverticulitis. *Ann Surg*. 1986; **204**: 128–32.
15 Townsend R R, Jefrey R B, Jr, Laing F C. Cecal diverticulitis differential from appendicitis using graded-compression sonography. *AJR*. 1989; **152**; 1229–30.
16 Parulekar S G. Sonography of colonic diverticulitis. *J Ultrasound Med*. 1985; **4**: 659–66.
17 Wilson R, Toi A. The value of sonography in the diagnosis of acute diverticulitis of the colon. *AJR*. 1990; **154**: 1199–202.
18 Boles, R S, Jr, Jordan S H. The clinical significance of diverticulosis. *Gastroenterology*. 1958; **35**: 579–582.
19 Haglund U, Hellberg C, Johnsén C, Hultén L. Complicated diverticular disease of the sigmoid colon. *Ann Chir Gynecol*. 1979; **68**: 41–6.
20 Colcock B P. Surgical management of complicated diverticulitis. *N Engl J Med*. 1958; **259**: 570–3.
21 Parks T G. Natural history of diverticular disease of the colon. *Clin Gastroenterol*. 1975; **4**: 53–69.
22 Krukowski Z H, Matheson N A. Emergency surgery for diverticular disease complicated by generalized and faecal peritonitis: a review. *Br J Surg*. 1984; **74**: 921–7.
23 Van Paaschen H, Goldman S, Magnusson I, Rieger Å. Surgical management of acute diverticulitis of the sigmoid colon. A ten year experience at one hospital. (Submitted for publication).
24 Rodkey G V, Welch C E. Colonic diverticular disease with surgical management. A study of 338 cases. *Surg Clin North Am*. 1974; **54**: 655–74.

25 Goligher J. Diverticulosis and diverticulitis of the colon. In *Surgery of the anus, rectum and colon* (5th edn.). London: Baillière Tindall, 1984, pp. 1083–1116.
26 Rodkey G V, Welch C E. Changing patterns in the surgical treatment of diverticular disease. *Ann Surg*. 1984; **200**: 466–78.
27 Finlay J G, Carter D C. A comparison of emergency resection and staged management in perforated diverticular disease. *Dis Colon Rectum*. 1987; **30**: 929–33.
28 Greif J M, Fried G, McScherry C K. Surgical treatment of perforated diverticulitis of the sigmoid colon. *Dis Colon Rectum*. 1980; **23**: 483–7.
29 Killingback M. Management of perforative diverticulitis. *Surg Clin North Am*. 1983; **63**(1): 110.
30 Bohnen J, Boulanger M, Meakins J L, McLean P H. Prognosis in generalized peritonitis. Relations of cause and risk factors. *Arch Surg*. 1983; **118**: 285–90.
31 Bakker F C, Hoitsma H F W, Den Otter A G. The Hartmann procedure. *Br J Surg*. 1982; **69**: 580–82.
32 Haas P A, Haas P G. A critical evaluation of the Hartmann procedure. *Am Surg*. 1988; **54**: 380–85.
33 Berry A R, Turner W H, Mortensen N J, Kettlewell M G W. Emergency surgery for complicated diverticular disease. *Dis Colon Rectum*. 1989; **32**: 849–54.
34 Lambert M F, Knox R A, Schofield P F, Hancock B D. Management of the septic complications of diverticular disease. *Br J Surg*. 1986; **73**: 576–9.
35 Hinchey E J, Schaal P G, Richards G K. Treatment of perforated diverticular disease of the colon. *Adv Surg*. 1978; **12**: 85–109.

8

Experimental carcinogenesis at large bowel suture lines

Jane L. McCue and Robin K.S. Phillips

Introduction

The importance of colorectal carcinoma in the Western world cannot be overstated. The disease is the second most common cause of death due to cancer and is exceeded only by carcinoma of the bronchus in men and carcinoma of the breast in women. In the UK approximately 23 000 cases of colorectal carcinoma are diagnosed *per annum*[1] and around 17 000 deaths ensue.[2]

Although surgery is the main form of treatment for colorectal carcinoma, 'curative' procedures are only performed in approximately 60 per cent of patients, because of the advanced stage of disease at presentation in the remainder.[3–5] The anticipated 5-year survival rate for those who have undergone ostensibly 'curative' procedures is in the region of 50 per cent[6,7] and is explained by the presence of occult hepatic micrometastases in about 30 per cent of those who are thought to have no residual disease after surgery.[8]

Over the past 30 years operative mortality following surgery for colorectal carcinoma has declined[9] and is now is the region of 6 per cent,[10,11] otherwise, however, improvement of long-term survival rates has been disappointing. During this period there has been a general trend towards restorative surgery, particularly for middle rectal cancers. Whereas there is no doubt that restorative procedures result in an improved quality of life for the patient, the speculation regarding increased local recurrence rates continues.[5,12,13]

A variety of avenues exist whereby future long-term survival rates might be improved, the most important being the development of effective systemic therapy for disseminated colorectal carcinoma. For the present, survival improvements are likely to arise through earlier detection of the disease via screening programmes, improved operative mortality rates and a reduction in subsequent treatment failure due to local recurrence or distant metastases, by effective adjuvant therapy or modification of current technique.

The purpose of this chapter is to consider the experimental evidence that relates to enhanced anastomotic carcinogenesis in an attempt to develop measures that may redress this potential risk.

Anastomotic recurrence: the magnitude of the problem

Local recurrence remains an important problem following 'curative' surgery for colorectal carcinoma. The reported clinical incidence (mainly symptomatic disease) is in the region of 14 per cent for rectal cancer[14–18] and may be just as high for colonic carcinoma.[5]

However, a policy of planned 'second-look' laparotomy[19,20] and routine post-mortem studies[21] have confirmed that local recurrence is consistently underestimated by clinical means. Whereas in many cases locally recurrent disease is almost incidental to advanced disseminated disease, in others it represents a major cause of morbidity and mortality.[15,16]

Approximately 80 per cent of cases will have presented within 2 years of surgery[22,23] although over 5 per cent present after 5 years.[15] The tumour bed is the most common site for local recurrence, with only around a fifth of the cases developing in the regional nodes.[24] Even following 'curative' re-resection 5-year survival is low, between 20–30 per cent[23,25,26] whereas the majority of patients die within a few months.[25] The reported incidence of truly anastomotic recurrence varies from 1.8–25 per cent[28] although it is probably in the region of 4–5 per cent.[29–32] The variation in estimates probably reflects a difference in definition. The demonstration of recurrent tumour at an anastomosis is by no means synonymous with isolated 'anastomotic' recurrence because in many cases it represents recurrent disease that has grown in from the pelvis.

Pathogenesis of anastomotic recurrence

i) Incomplete excision

The major reason for local recurrence after ostensibly 'curative' surgery is incomplete tumour clearance. There is little doubt that improved training in pelvic dissection to minimize inadequate clearance or tumour disruption will, in time, lead to a reduction in local recurrence rates. Nonetheless local recurrence may still occur following excision of Dukes' A tumours[5,18,33,34] and intraperitoneal colonic tumours[5,24] in which resection margins should not be a problem. Further explanations for local recurrence seem warranted.

ii) Implantation metastases

The concept that exfoliated cancer cells are capable of generating implantation metastases dates back over a century.[35] Cases of metastatic deposits arising on raw surfaces, for example fissures or haemorrhoidectomy wounds are well recognized[36] and supported by experimental data. In a rabbit model, tumour cells were injected into a segment of gut followed by enterotomy and suture. This resulted in suture line recurrences in 12 out of 15 animals[37] and confirmation that sutures could transfer carcinoma cells to other sites followed.[38] More recently it has been demonstrated that malignant cells adhere to suture materials.[39] Further work in our laboratory has shown that tumour cells were least adherent to monofilament steel, nylon and polydioxanone sutures and most adherent to protein based and multifilament sutures (C R Uff, unpublished data).

Patterns of exfoliated cell dissemination were studied by McGrew who demonstrated that in half of the resection specimens studied exfoliated cancer cells were found at least 10–15 cm from the tumour, either proximal or distal.[40] Seemingly reassuring experimental evidence was presented by Rosenberg who demonstrated that cells exfoliated from colorectal cancers were not viable and thus an unlikely cause of anastomotic recurrence.[41] These findings were later refuted by Umpleby *et al.* who found millions of viable cancer cells in the bowel lumen adjacent to the site of bowel resection in up to 70 per cent of cases examined.[42] Subsequent work has revealed that these exfoliated cells maintain their proliferative potential in immuno-suppressed mice.[43] Implantation metastases seem a likely explanation for some cases of anastomotic recurrence and operative measures should be taken either to kill the liberated cells or limit their spread. Whilst the efficacy of occlusive tapes around the colon has been challenged, *in vitro* testing of chlorhexidine-cetrimide and povidone-iodine has shown rapid cytotoxicity for cancer cells.[44] Rectal irrigation and preparation of the transected bowel agent with these agents is therefore recommended.

iii) Metachronous carcinoma

Clinical evidence

It has been suggested that some cases of anastomotic recurrence, particularly the 30 per cent occurring more than 2 years after operation,[45] may represent metachronous carcinomas at the anastomosis. It is believed that the unstable mucosa evoked by reparative hyperplasia in the vicinity of an anastomosis may facilitate carcinogenesis.

Good evidence exists that colorectal carcinoma is a multifocal disease. 20 per cent of patients with colorectal neoplasia have multiple tumours.[46] Synchronous carcinomas occur in approximately 3–4 per cent of patients[47] and a similar incidence is reported for subsequent metachronous carcinomas.[30,48] Moreover if simultaneous adenomas are noted in the carcinoma resection specimen, the risk of a second cancer rises steeply reaching 10 per cent at 25 years.[46] In addition, conditions such as familial adenomatous polyposis and chronic ulcerative colitis, which have manifest proliferative abnormalities and a strong association with carcinoma, are both associated with an elevated risk of multiple tumours.[49]

Further support for this theory is the demonstration of 'field change' within the morphologically normal colon of patients with colorectal carcinoma that manifests itself as abnormalities of mucin histochemistry and altered cytokinetics. Filipe and Branfoot have shown replacement of goblet cell sulphomucins by sialomucins over a varying distance of mucosa adjacent to carcinoma specimens.[50] This represents a return to the situation found in fetal gut. Furthermore, an association between sialomucin predominance at the resection margin and local recurrence has been demonstrated.[51]

A colonoscopic study has investigated anastomotic histological findings in 28 patients who had undergone resection and anastomosis for carcinoma up to 2 years previously.[52] Inflammation was noted macroscopically in five and microscopically in six others. Additional cryptal abnormalities were present and sialomucin predominance, which had not been observed at the time of operation, was seen in seven.

Experimental evidence

a) Non-specific enhancement by trauma

The association between trauma, chronic irritation and carcinogenesis has long been recognized. Although few investigators have concentrated solely on the colonic effect of non-specific trauma, it is appropriate to present data obtained in other organs and from which inference may be drawn.

In 1927 Deelman applied tar to mouse skin to produce papillomas, and subsequently made incisions in macroscopically unaffected areas.[53] The ensuing tumour formation was most extensive in the immediate vicinity of the scar. 7 years later Orr induced subcutaneous fibrosis in mice by temporary implantation of linen sutures. When tar was applied to the scarred area, tumour induction occurred more rapidly than in adjacent areas.[54] The role of injury was further emphasized by Rous and Kidd[55] who demonstrated that application of mechanical trauma to rabbit ears, which had been painted with tar, caused the reappearance of previously regressed tumours. Multiple skin excisions in mice treated with benzpyrene resulted in a tumour yield up to three times higher than following a single excision, which led to the postulate that the underlying cause was enhanced cellular proliferation associated with wound healing.[56]

Gottfried *et al.* explored the effect of repeated surgical trauma, commencing 6 weeks after dibenzpyrene administration. In addition to the carcinogen, animals received wounding or laparotomy three times weekly. By 14 weeks the incidence of tumours in the laparotomy group was 82 per cent, compared with 65 per cent in the wounding group and 55 per cent in the carcinogen control group. There was a concomitant increase in large, aggressive tumours in the laparotomy group.[57] Thus a quantitative relation existed between trauma severity and reduction in tumour latency. Recent work has confirmed that operations facilitate tumour growth.[58] Rats were given 10^7 colorectal cancer cells intraperitoneally either by injection or at laparotomy. At 26 weeks tumours had developed in 89 per cent of the laparotomy group compared with only 49 per cent of the injection group ($p<0.001$). Repeated surgery further enhanced cancer yield.

From these results it can be concluded that surgical trauma enhances experimental carcinogenesis, even when the trauma occurs at a distance from the tumour site. Although in most studies carcinogen administration has preceded trauma, previous injury may exert an effect through scarring.[54]

b) Compensatory intestinal hyperplasia

The majority of work in this field has been carried out by Williamson and co-workers who have performed a series of experiments probing the influence of various intestinal resections on colorectal carcinogenesis. Initial work explored the effect of a 50 per cent small bowel resection on azoxymethane-induced carcinogenesis.[59] A significant increase in mean number of colon tumours per animal from 1.6 in the control group to 2.9 in the resection group ($p<0.02$) was seen. Carcinogenic potential was also enhanced by ileal bypass, although to a lesser degree than resection.[60,61] It appeared that loss of functioning ileum enhanced experimental colorectal carcinogenesis, which was mediated by compensatory hyperplasia.[62]

In contrast to the marked effects of small bowel resection on colorectal carcinogenesis, the influence of colonic resection was limited and site-dependent. The results of subtotal colectomy and ileorectal anastomosis on rectal carcinogenesis were conflicting; Williamson showed a modest increase in susceptibility,[63] whilst others found a decrease[64] or little change.[64] Following limited transverse colectomy there was a marked tendency for tumours to cluster around the anastomosis although no overall change in colonic tumour yield occurred.[66,67] Williamson *et al.* also noted this tendency following caecal resection, and right or left hemicolectomy, but found no enhancement of carcinogenesis at other colonic sites, despite the fact that left hemicolectomy induced right colonic hyperplasia, which might be expected to enhance tumour yield in that region.[68] Although tumours developed preferentially around a colonic anastomosis, anastomoses to the caecum or proximal colon seemed relatively protected from this phenomenon.[68,69]

c) Colonic injury

Generalized mucosal colonic injury, caused by infection with *Citrobacter freundii*[70] or acetic acid colitis,[71] enhance chemical carcinogenesis.

The predisposition for tumours to develop around colonic anastomoses, discussed above, was independent of bowel resection. It has been seen with both locally applied and systemic carcinogens, given pre- or postoperatively, provided at least one limb of the anastomosis was a common site for tumour formation. Experimentally it has been documented after colon transection or colotomy and repair[72–77] and following transection with redirection of the gastrointestinal tract.[77–82] Despite complete re-epithelialization and a return to normal histological appearances by 12 weeks the anastomosis remained more susceptible to carcinogenesis.[74]

Early work with 3,2-dimethyl-4-aminobiphenyl showed that tumours were prone to develop at colostomy sites.[83] This has since been confirmed with 1,2-dimethylhydrazine.[84,85] Indeed a strain of rat exists in which the stimulus of colostomy formation alone is sufficient to produce stomal tumours. Even handling the colon at laparotomy has been shown to increase the yield of carcinogen-induced tumours in the manipulated bowel.[86] Interestingly ureterosigmoidostomy, which has a greatly increased risk of colorectal cancer in man, has been enough to produce anastomotic tumours in five out of eight rats without carcinogen.[80]

Accelerated tumour formation has been observed at suture lines. Rats, which have earlier received 12 weekly doses of 1,2-dimethylhydrazine before colonic transection, developed carcinoma *in situ* at the anastomosis, but not at other sites, within 2 weeks of operation.[72] Enhanced carcinogenesis at anastomoses and in areas of marked faecal contact[81,82] has led to the suggestion that the underlying common factor was prolonged exposure to faecal carcinogen. In the experimental situation this seems unlikely on two counts; firstly, less than 1 per cent of 1,2-dimethylhydrazine is excreted in bile; and secondly, no anastomotic stenoses were reported.[81] A more plausible unifying explanation is through an effect on colonic cellular proliferation which will be considered later.

Several investigators have explored the role of sutures on carcinogenesis. Pozharisski devised a technique whereby a suture was placed in the caecum to create a diverticulum. Within 7–10 days the suture partially cut through to encircle an area of necrotic mucosa. 87 per cent of animals that commenced 1,2-dimethylhydrazine 1 week later developed caecal tumours compared with just 23 per cent in the unoperated controls.[87] The majority of tumours (27 of 43) developed around the suture, and this tendency persisted even when carcinogen exposure was delayed for 2 months after operation, by which time the mucosa had healed. McGregor confirmed that sutures can enhance tumour risk; tumour yield being similar following suture implantation alone and when there was colotomy with suturing. This implies either that sutures are the sole influential component, or if sutures and colotomy separately influence carcinogenesis, the effects are not additive.[75]

Calderisi placed a variety of suture materials into the caecum of rats that had already been exposed to carcinogen. A significant increase in caecal tumour yield was found with Dexon, Vicryl, silk or multifilament steel compared with either control or catgut sutured animals.[88] In addition, the steel group developed more tumours in the descending colon, which seemed to correlate with beta-glucuronidase activity.

Both Phillips and McGregor have addressed the influence of different suture materials on carcinogenesis at a sutured colotomy.[73,75] Steel sutures were common to both groups, to simulate the effect of staples used in anastomotic guns. At first glance the results would appear to be directly conflicting. Phillips found that 10 out of 16 (63 per cent) colonic tumours arose at the anastomosis in the steel sutured group compared with 2 out of 12 (17 per cent) in the silk sutured group ($p = 0.019$). In contrast, McGregor found only 17 per cent of tumours were anastomotic in the steel group compared with 53 per cent in the polyamide group and 58 per cent in the polyglycolic acid group. However, it is conceivable that these results can be explained on the basis of differences in experimental design. Whereas Phillips allowed 8 weeks to elapse **after** operation, to permit any inflammatory response to settle, before commencing carcinogen exposure, McGregor treated his animals with azoxymethane **before** surgery. Consequently the sutures were acting on different phases of the carcinogenic process in each study.

Enhanced anastomotic tumour risk has been attributed to the persistence of suture material acting as a chronic irritant, although the validity of this theory is arguable. Anastomotic tumours developed in 7 out of 17 (41 per cent) with persistent sutures compared with five out of 47 without (11 per cent) in the work by Phillips. He also noted that scarring was more severe in those sutured with steel and most severe in those with persisting sutures; there was little difference noted with regard to inflammation. In contrast others reported a lower anastomotic tumour yield in those with persisting sutures.[75,76]

Technique of suture insertion also appears important. Whilst consistently placing the suture in one direction, for example from mucosa out or from serosa in, had no effect on tumour yield,[73] the proportion of tumours occurring at the anastomosis was significantly greater using transmural sutures (16 out of 22) compared with seromuscular sutures (five out of 13, $p<0.05$).[76] The crucial factor would seem to be interaction of the suture material with the mucosa.

In order to clarify some of these factors we have recently carried out a study that employed a novel technique of 'sutureless' colonic anastomosis to explore the relative role of sutures and the healing colonic wound in experimental colorectal carcinogenesis (see Fig. 8.1). 166 male F344 rats received 12 consecutive weekly subcutaneous injections of azoxymethane 10 mg/kg/week. 77 animals underwent operation 8 weeks before carcinogen treatment and 89 received carcinogen prior to operation. Either: 1) a sham procedure; or a 5 mm transverse colotomy repaired with 2) 4 interrupted 5/0 sutures of either a) silk, b) polyglactin (Vicryl), c) stainless steel or 3) a 'sutureless' technique were performed. Animals were killed 28 weeks after the first dose of carcinogen.

Animals with anastomotic tumour were found in 46 per cent of the sham group, 41 per cent of the 'sutureless group' and 68 per cent in the sutured group ($p<0.05$ versus sham $X^2 = 4.4$; $p<0.02$ versus 'sutureless', $X^2 = 6.35$). The corresponding figures for anastomotic cancer were 9, 22 and 38 per cent ($p<0.002$ versus sham, Fisher's exact test).

Overall no significant differences in tumour yield were noted between silk, Vicryl and steel. Moreover several differences were noted between the two carcinogen models, confirming our hypothesis that this was the reason for previous conflicting results. The suture which emerged as safest in the preoperative carcinogen group (Steel) was least safe in the postoperative carcinogen group. All sutures seem to enhance anastomotic tumour formation, over and above the enhancement caused by a healing enterotomy, and we suggest that a 'sutureless' anastomosis may diminish this risk.[89]

Cellular proliferation at anastomoses

i) Clinical evidence

It is worth examining the evidence to support the theory that enhanced anastomotic carcinogenesis is due to alteration in cellular proliferation.

Matthews *et al.* calculated the proliferative index of cells obtained by cytological brushings from the anastomosis in ten patients who had undergone partial colonic resection. The mean anastomotic proliferative index was 6.2 ± 1.8, which was significantly greater than the control value 2.2 ± 0.3

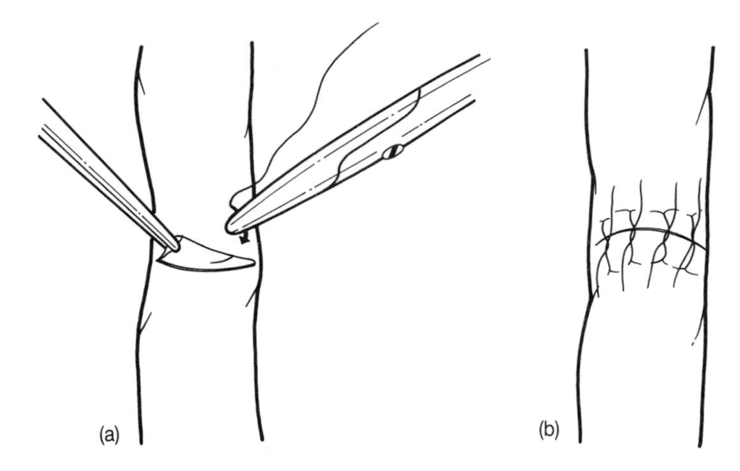

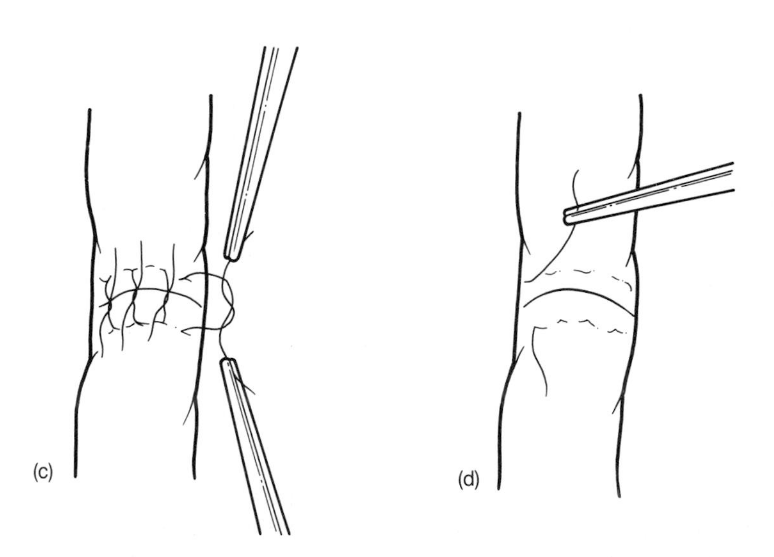

Fig. 8.1 **(a)** Transverse enterotomy at start of suturing **(b)** four sutures inserted, one throw placed on each and ends left long **(c)** 1-hour later sutures gently disentangled and removed **(d)** after suture removal bowel integrity maintained by fibrin seal.

($p<0.001$). Four of the patients had indices above the normal range and of these two developed recurrences. The shift of proliferative activity to the upper crypt seen here may highlight those patients who are at greatest risk of local recurrence.

Indirect support is obtained from the fact that similar cell proliferation abnormalities are seen in a variety of high-risk cancer groups, i.e. carcinoma and adenoma patients,[90] those with a genetic predisposition to colorectal carcinoma,[91] patients with inflammatory bowel disease,[92,93] the elderly[94] and women after pelvic irradiation.[95]

Table 8.1 Tumour yields with different anastomotic techniques

	n	Anastomotic tumours: cancer	Proportion of tumours at anastomosis
Sham	35	19:3	0.23
'Sutureless'	32	14:8	0.47
Sutured	99	85:46	0.55

ii) Experimental evidence

In experimental animals there is a marked hyperplastic response to carcinogen treatment. As in the clinical situation there is often a close relation between proliferation status and susceptibility to carcinogenesis; thus baseline colonic cellular proliferation is highest in carcinogen-sensitive animals.[96]

Predictably, various other hyperplastic stimuli also enhance carcinogenesis. When mice have been infected with a variant of *Citrobacter freundii*, during 1,2-dimethylhydrazine treatment there has been a reduction in the latent period before tumours emerge but no influence on established tumours.[70]

Small bowel resection and bypass produce a rapid increase in colonic proliferation and both have been linked with an enhancement of subsequent colon carcinogenesis.[59,61] When operation precedes carcinogen administration the major effect has been reduction in the latent period before tumour development. Conversely carcinogenesis is diminished in defunctioned colon with its associated mucosal hypoplasia.[84,97]

Dietary supplementation with 15 per cent cellulose during carcinogen treatment produces a 35 per cent fall in mitotic index compared to controls, and a significant reduction in tumour incidence from 88 per cent to 38 per cent.[98] Conversely a diet containing 10 per cent guar produces an elevation in labelling index to 12.2 per cent versus 7.9 per cent in controls and a doubling of colonic tumours ($p<0.05$).[99]

Agents such as beta-sitosterol,[100] butylated hydroxyanisole[101] and ascorbic acid,[102] which depress colonic cellular proliferation, also decrease tumourigenesis.

In conclusion, elevated cellular proliferation during initiation appears to increase the prevalence of tumour formation and to reduce the latent phase, particularly in the distal colon. There are two underlying mechanisms: firstly, there is less time for repair of genetic mutations, which thus become 'fixed' by mitosis; and secondly a larger number of cells are exposed to carcinogenic stimuli at any one time, which thereby increases the chance of alteration to the genome.

As far as promotion is concerned, the situation is more complex and as yet unresolved. Although some investigators have shown that elevated proliferative activity during promotion enhances carcinogenesis,[59] others have found exactly the opposite.[103] Furthermore, the duration of promotion is much longer than initiation and may well involve several stages. It is possible that cell kinetic changes could exert differing effects at separate times throughout the phase of promotion. Slowing of cellular production might allow an initiated cell to establish itself within the crypt, instead of being transported rapidly to the surface and shed.

A number of workers have specifically investigated anastomotic cellular proliferation. Elevated proliferation was first shown by Kanazawa in 1975.[79] The labelling index increased from 6.7 per cent in adjacent descending colon to 11.6 per cent at the anastomosis 25 weeks postoperatively in animals that had received topical N-methyl-N'-nitro-N-nitrosoguanidine. Barkla subsequently studied the duration of the enhanced proliferative response in rats after transverse colon transection and monofilament nylon anastomosis, with or without carcinogens.[72] A significant elevation in the mitotic index at the anastomosis was seen for at least 72 days in those animals that received carcinogens whereas in those that did not, the response took longer to achieve and was briefer. Further work by Roe quantified these changes in rats given azoxymethane preoperatively.[74] At the anastomosis, crypt height increased by a maximum of 48 per cent at 4 weeks postoperatively and returned to normal by 12 weeks. Labelling index also peaked at 4 weeks with a figure of 26 per cent but by 12 weeks was still elevated at 14 per cent compared with the control value, 10 per cent. Proliferative changes thus persist even when re-epithelialization is complete. Appleton *et al*. demonstrated that anastomotic crypt cell production rate (CCPR) was increased by a factor of 2.6 over intact colon 7 weeks after construction, although defunction by transverse colostomy cancelled the effect.[77]

McGregor confirmed that carcinogen and anastomosis both additively enhance CCPR over a period of 12 weeks. Simple suture placement produced a similar degree of elevation to colotomy plus resuture; subsequent tumour yield was similar in both groups and no obvious differences emerged between the sutures tested.[75] In contrast, Matthews found a hyperplastic response at the anastomosis after 1 week and a return to normality thereafter. Transmural suturing was associated with a significantly greater elevation in the bromodeoxyuridine labelling index than seromuscular closure at 26 weeks,[26] but no correlation was seen with suture persistence. Caecal suturing resulted in increased activity deep in the mucosa in an area abutting the suture even at 149 days, although generally labelling was normal by 60 days.[87] It is possible that part of the proliferative response seen at anastomoses is mediated by ischaemia, which is known to reduce cell cycle duration in the colon.

Clearly cellular proliferation is heightened around an anastomosis and this may persist beyond the period when such a response is useful. Sutures alone seem to possess the ability to evoke this reaction but it is unknown whether proliferation would be lower at a 'sutureless' anastomosis.

In an attempt to explain our experimental finding of diminished carcinogenesis at 'sutureless' compared with sutured anastomoses we explored the influence of different techniques on anastomotic crypt cell production rate.

A further 80 male F344 rats were used. A 5 mm transverse colotomy was created, which was repaired with: four interrupted 5/0 sutures of silk, stainless steel or polyglactin 910 (Vicryl) or a 'sutureless' closure. Five animals in each group were killed after 1 week, 1 month, 3 months or 6 months. CCPR was assessed by the stathmokinetic technique. At each time interval the number of metaphase figures in 20 crypts were counted and plotted against time. The CCPR was derived from the slope of the regression line.

In the sutured animals anastomotic CCPR was significantly greater than adjacent descending colon CCPR for at least 3 months postoperatively (see Table 8.2). Differences between individual sutures were most marked at 1 month (anastomotic CCPR: silk = 15.0, steel = 9.7 [$f = 10$, $p<0.005$]; Vicryl = 7.1 [versus silk $f = 18.1$, $p<0.005$]).

By contrast there was no significant elevation of the CCPR at the 'sutureless' anastomosis compared with the intact colon at any time point. These results lend support to our hypothesis that the increased incidence of experimental tumours at sutured compared to 'sutureless' anastomoses is mediated via an enhanced and prolonged elevation of anastomotic cellular proliferation by sutures.

Table 8.2 Cellular proliferation in sutured animals

CCPR (cells/crypt/hour)	Anastomosis	Descending colon
1 week	10.3	7.8*
1 month	10.6	7.9**
3 months	8.7	7.7†
6 months	7.9	8.8

*$f = 9.48$, $p < 0.005$; **$f = 10.3$, $p < 0.005$; †$f = 4.4$, $p < 0.05$.

Clinical implications

We have seen clearly that the suture component of an anastomosis enhances carcinogenesis over and above the enhancement of a healing enterotomy. This is an important point when considering methods of prevention because healing bowel ends cannot be avoided in restorative surgery, whereas there are other ways of joining bowel together that avoid the use of implanted suture materials or staples. Recently there has been a resurgence of interest in sutureless intestinal anastomotic techniques. A variety of devices, for example, the AKA instrument[104,105] and the biofragmentable anastomotic ring[106–109] are now available and utilize the principle of compression to unite bowel ends. The initial clinical results are comparable to conventional sutured and stapled intestinal anastomoses. Use of these techniques should minimize the risk of anastomotic carcinogenesis and other theoretical advantages exist that should allow these devices to become widely used.

Other 'sutureless' anastomotic techniques have also been explored experimentally. Glued anastomoses appear unsafe[110,111] but laser-welded anastomoses seem highly promising in experimental studies and further development of this technique is warranted.[112]

Appleton *et al.* have suggested that a transverse colostomy raised during an anterior resection may lower the incidence of metachronous carcinoma development at the colorectal anastomosis. In addition, they advised that closure should be delayed for several months until the anastomosis is cytokinetically stable.[77]

An alternative approach might be to explore dietary manipulation or calcium supplementation to modify anastomotic cellular proliferation, in those patients who have undergone sutured or stapled colorectal anastomoses.

References

1 Office of Population Censuses and Surveys, 1985.
2 Office of Population Censuses and Surveys, 1990.
3 Clarke D N, Jones P F, Needham C D. Outcome in colorectal carcinoma: seven-year study of a population. *Br Med J*. 1980; **280**: 43–5.
4 Umpleby H C, Bristol J B, Rainey J B, Williamson R C. Survival of 727 patients with single carcinomas of the large bowel. *Dis Colon Rectum*. 1984; **27**: 803–10.
5 Phillips R K, Hittinger R, Blesovsky L, *et al.*. Local recurrence following 'curative' surgery for large bowel cancer: I. The overall picture. *Br J Surg*. 1984; **71**: 12–16.
6 Shepherd J M, Jones J S. Adenocarcinoma of the large bowel. *Br J Cancer*. 1971; **25**: 680–90.
7 Goligher J C. *Surgery of the anus, rectum and colon* (5th edn). London: Baillière Tindall, 1984, pp. 547.

8 Finlay I G, Meek D R, Gray H W, *et al.*. Incidence and detection of occult hepatic metastases in colorectal carcinoma. *Br Med J Clin Res*. 1982; **284**: 803–5.
9 Ohman U. Colorectal carcinoma – trends and results over a 30-year period. *Dis Colon Rectum*. 1982; **25**: 431–40.
10 Malmberg M, Graffner H, Ling L, Olsson S A. Recurrence and survival after anterior resection of the rectum using the end to end anastomotic stapler. *Surg Gynecol Obstet*. 1986; **163**: 231–4.
11 Brown S C, Abraham J S, Walsh S, Sykes P A. Risk factors and operative mortality in surgery for colorectal cancer. *Ann R Coll Surg Engl*. 1991; **73**: 269–72.
12 Wolmark N, Fisher B. An analysis of survival and treatment failure following abdominoperineal and sphincter-saving resection in Dukes' B and C rectal carcinoma. A report of the NSABP clinical trials. National Surgical Adjuvant Breast and Bowel Project. *Ann Surg*. 1986; **204**: 480–89.
13 Neville R, Fielding L P, Amendola C. Local tumor recurrence after curative resection for rectal cancer. A ten-hospital review. *Dis Colon Rectum*. 1987; **30**: 12–17.
14 Goligher J C, Dukes C E, Bussey H J. Local recurrences after sphincter-saving excisions for carcinoma of the rectum and rectosigmoid. *Br J Surg*. 1951; **39**: 199–211.
15 Cass A W, Million R R, Pfaff W W. Patterns of recurrence following surgery alone for adenocarcinoma of the colon and rectum. *Cancer*. 1976; **37**: 2861–5.
16 Rao A R, Kagan A R, Chan P M, *et al.*. Patterns of recurrence following curative resection alone for adenocarcinoma of the rectum and sigmoid colon. *Cancer*. 1981; **48**: 1492–5.
17 Phillips R K, Hittinger R, Blesovsky L, *et al*. Local recurrence following 'curative' surgery for large bowel cancer: II. The rectum and rectosigmoid. *Br J Surg*. 1984; **71**: 17–20.
18 Feil W, Wunderlich M, Kovats E, *et al.*. Rectal cancer: factors influencing the development of local recurrence after radical anterior resection. *Int J Colorectal Dis*. 1988; **3**: 195–200.
19 Wangensteen O H, Lewis F J, Arhelger S W, *et al.*. An interim report on the 'second look' procedure for cancer of the stomach, colon and rectum and for 'limited intraperitoneal carcinosis'. *Surg Gynecol Obstet*. 1954; **99**: 257–67.
20 Gunderson L L, Sosin H. Areas of failure found at re-operation (second or symptomatic look) following 'curative surgery' for adenocarcinoma of the rectum. *Cancer*. 1974; **34**: 1278–92.
21 Gilbert J M, Jeffrey I I, Evans M, Kark A E. Sites of recurrent tumour after 'curative' colorectal surgery: implications of adjuvant therapy. *Br J Surg*. 1984; **71**: 203–5.
22 Tyndal E C, Dockerty M B, Waugh J M. Pelvic recurrence of carcinoma of the rectum. *Surg Gynecol Obstet*. 1964; **118**: 47–51.
23 Polk H C J, Spratt J S J. Recurrent colorectal carcinoma: detection, treatment, and other considerations. *Surgery*. 1971; **69**: 9–23.
24 Willett C, Tepper J E, Cohen A, *et al.*. Local failure following curative resection of colonic adenocarcinoma. *Int J Radiat Oncol Biol Phys*. 1984; **10**: 645–51.
25 Tonak J, Gall F P, Hermanek P, Hager T H. Incidence of local recurrence after curative operations for cancer of the rectum. *Aust N Z J Surg*. 1982; **52**: 23–7.

26 Fritsch A, Herbst F, Schiessel R. Local recurrence after colorectal cancer. *Wien Med Wochenschr*. 1988; **138**: 313–16.
27 Hardy K J, Cuthbertson A M, Hughes E S. Suture-line neoplastic recurrence following large-bowel resection. *Aust N Z J Surg*. 1971; **41**: 44–6.
28 Judd E S, Bellegie N J. Carcinoma of the rectosigmoid and upper part of the rectum. Recurrence following low anterior resection. *Arch Surg*. 1952; **64**: 697–706.
29 Labow S B, Salvati E P, Rubin R J. Suture-line recurrence in carcinoma of the colon and rectum. *Dis Colon Rectum*. 1975; **18**: 123–5.
30 Hughes E, McDermott F T, Johnson W R. Management of recurrent large bowel cancer. In DeCosse J J (Ed), *Large bowel cancer*. Edinburgh, London, Melbourne, New York: Churchill Livingstone, 1981, pp. 205–15.
31 Pihl E, Hughes E S, McDermott F T, Price A B. Recurrence of carcinoma of the colon and rectum at the anastomotic suture line. *Surg Gynecol Obstet*. 1981; **153**: 495–6.
32 Barkin J S, Cohen M E, Flaxman M, *et al.*. Value of a routine follow-up endoscopy program for the detection of recurrent colorectal carcinoma. *Am J Gastroenterol*. 1988; **83**: 1355–60.
33 Wheelock F C, Toll G, McKittrick L S. An evaluation of the anterior resection of the rectum and low sigmoid. *N Engl J Med*. 1959; **260**: 526–30.
34 Olson R M, Perencevich N P, Malcolm A W, *et al.*. Patterns of recurrence following curative resection of adenocarcinoma of the colon and rectum. *Cancer* 1980; **45**: 2969–74.
35 Gerster A G. On surgical dissemination of cancer. *N Y Med J*. 1885; **41**: 233–6.
36 Goligher J C. *Surgery of the anus, rectum and colon* (5th edn). London: Ballière Tindall, 1984, pp. 454–5.
37 Vink M. Local recurrence of cancer in the large bowel: the role of implantation metastases and bowel disinfection. *Br J Surg*. 1954; **41**: 431–3.
38 Haverback C Z, Smith R R. Transplantation of tumour by suture thread and its prevention. An experimental study. *Cancer*. 1959; **12**: 1029–42.
39 O'Dwyer P, Ravikumar T S, Steele G. Serum dependent variability in the adherence of tumour cells to surgical sutures. *Br J Surg*. 1985; **72**: 466–9.
40 McGrew E A, Laws J F, Cole W H. Free malignant cells in relation to recurrence of carcinoma of the colon. *JAMA*. 1954; **154**: 1251–4.
41 Rosenberg I L, Russell C W, Giles G R. Cell viability studies on the exfoliated colonic cancer cell. *Br J Surg*. 1978; **65**: 188–90.
42 Umpleby H C, Fermor B, Symes M O, Williamson R C. Viability of exfoliated colorectal carcinoma cells. *Br J Surg*. 1984; **71**: 659–63.
43 Fermor B, Umpleby H C, Lever J V, *et al.*. The proliferative and metastatic potential of exfoliated colorectal carcinoma cells. *J Natl Cancer Inst*. 1985; **74**: 1161–8.
44 Umpleby H C, Williamson R C. The efficacy of agents employed to prevent anastomotic recurrence in colorectal carcinoma. *Ann R Coll Surg Engl*. 1984; **66**: 192–4.
45 Umpleby H C, Williamson R C. Anastomotic recurrence in large bowel cancer. *Br J Surg*. 1987; **74**: 873–8.
46 Muto T, Bussey H J, Morson B C. The evolution of cancer of the colon and rectum. *Cancer*. 1975; **36**: 2251–70.
47 Finan P J, Ritchie J K, Hawley P R. Synchronous and 'early metachronous' carcinomas of the colon and rectum. *Br J Surg*. 1987; **74**: 945–7.
48 Bussey H J, Wallace M H, Morson B C. Metachronous carcinoma of the large intestine and intestinal polyps. *Proc R Soc Med*. 1967; **60**: 208–10.

49 Greenstein A J, Slater G, Heimann T M, *et al.*. A comparison of multiple synchronous colorectal cancer in ulcerative colitis, familial polyposis coli, and de novo cancer. *Ann Surg*. 1986; **203**: 123–8.
50 Filipe M I, Branfoot A C. Abnormal patterns of mucus secretion in apparently normal mucosa of large intestine with carcinoma. *Cancer*. 1974; **34**: 282–90.
51 Dawson P M, Habib N A, Rees H C, *et al.*. Influence of sialomucin at the resection margin on local tumour recurrence and survival of patients with colorectal cancer: a multivariate analysis. *Br J Surg*. 1987; **74**: 366–9.
52 Sunter J P, Higgs M J, Cowan W K. Mucosal abnormalities at the anastomosis site in patients who have had intestinal resection for colonic cancer. *J Clin Pathol*. 1985; **38**: 385–9.
53 Deelman H T. The part played by injury and repair in the development of cancer. *Br Med J*. 1927; **1**: 872.
54 Orr J W. The influence of ischaemia on the development of tumours. *Br J Exp Pathol*. 1934; **15**: 73–9.
55 Rous P, Kidd J G. Conditional neoplasms and subthreshold neoplastic states; study of tar tumours of rabbits. *J Exp Med*. 1941; **73**: 365–90.
56 Pullinger B D. A measure of the stimulating effect of simple injury combined with carcinogenic chemicals on tumour formation in mice. *J Pathol Bact*. 1945; **57**: 477–81.
57 Gottfried B, Molomut N, Patti J. Effect of repeated surgical trauma on chemical carcinogenesis. *Cancer Res*. 1961; **21**: 658–60.
58 Weese J L, Ottery F D, Emoto S E. Do operations facilitate tumor growth? An experimental model in rats. *Surgery*. 1986; **100**: 273–7.
59 Williamson R C, Bauer F L, Oscarson J E, *et al.*. Promotion of azoxymethane-induced colonic neoplasia by resection of the proximal small bowel. *Cancer Res*. 1978; **38**: 3212–7.
60 Scudamore C H, Freeman H J. Effects of small bowel transection, resection, or bypass in 1,2-dimethylhydrazine-induced rat intestinal neoplasia. *Gastroenterology*. 1983; **84**: 725–31.
61 Rainey J B, Davies P W, Williamson R C. Relative effects of ileal resection and bypass on intestinal adaptation and carcinogenesis. *Br J Surg*. 1984; **71**: 197–202.
62 Williamson R C. Intestinal adaptation: factors that influence morphology. *Scand J Gastroenterol Suppl*. 1982; **74**: 21–9.
63 Williamson R C N. Experimental aspects. In *Topics in gastroenterology*, 1991, pp. 265–84.
64 Ferulano G P, Cruse J P, Lewin M R, Clark C G. Subtotal colectomy in the dimethylhydrazine-treated rat. A surgical model of colorectal cancer. *Eur Surg Res*. 1982; **14**: 393–400.
65 Celik C, Mittelman A, Lewis D, *et al.*. Effect of colectomy on carcinogenicity of symmetrical dimethylhydrazine in rats. *Surgical Forum*. 1980; **31**: 415–7.
66 Rubio C A, Nylander G. Surgical resection of the rat colon: effects on carcinogenesis by 1,2-dimethylhydrazine. *J Nat Cancer Inst*. 1982; **68**: 813–5.
67 Rubio C A, Nylander G, Wallin B, *et al.*. Partial colon resection as a promotor of cancer growth in the rat. *J Surg Oncol*. 1984; **27**: 236–8.
68 Williamson R C, Davies P W, Bristol J B, Wells M. Intestinal adaptation and experimental carcinogenesis after partial colectomy. Increased tumour yields are confined to the anastomosis. *Gut*. 1982; **23**: 316–25.
69 Werner B, Heer K, Mitschke H, *et al.*. Experimental carcinogenesis in the resected colon of the rat. *Z Krebsforsch*. 1977; **89**: 53–60.

70 Barthold S W, Jonas A M. Morphogenesis of early 1,2-dimethylhydrazine-induced lesions and latent period reduction of colon carcinogenesis in mice by a variant of *Citrobacter freundii*. *Cancer Res*. 1977; **37**: 4352–60.
71 Hagihara P F. Experimental colitis as a promoter in large-bowel tumorigenesis. *Arch Surg*. 1982; **117**: 1304–307.
72 Barkla D H, Tutton P M. The influence of surgical transection and anastomosis on the rate of cell proliferation in the colonic epithelium of normal and DMH-treated rats. *Carcinogenesis*. 1983; **4**: 1323–55.
73 Phillips R K, Cook H T. Effect of steel wire sutures on the incidence of chemically induced rodent colonic tumours. *Br J Surg*. 1986; **73**: 671–4.
74 Roe R, Fermor B, Williamson R C. Proliferative instability and experimental carcinogenesis at colonic anastomoses. *Gut*. 1987; **28**: 808–15.
75 McGregor J R. *Clinical and experimental studies of gastro-intestinal anastomotic techniques*. University of Glasgow: MD Thesis, 1988.
76 O'Donnell A F, O'Connell P R, Royston D, *et al*.. Suture technique affects perianastomotic colonic crypt cell production and tumour formation. *Br J Surg*. 1991; **78**: 671–4.
77 Appleton G V, Davies P W, Williamson R C. Effect of defunction on cytokinetics and cancer at colonic suture lines. *Br J Surg*. 1990; **77**: 768–72.
78 Gennaro A R, Villanueva R, Sukonthaman Y, *et al*.. Chemical carcinogenesis in transposed intestinal segments. *Cancer Res*. 1973; **33**: 536–41.
79 Kanazawa K, Yamamoto T, Sato S. Experimental induction of colonic carcinomas in rats. Analysis of factors influencing upon the incidence. *Jpn J Exp Med*. 1975; **45**: 439–56.
80 Steele G, Jr, Crissey M, Gittes R, *et al*.. Potentiation of dimethylhydrazine bowel carcinogenesis in rats. *Cancer*. 1981; **47**: 2218–21.
81 Filipe M I, Scurr J H, Ellis H. Effects of fecal stream on experimental colorectal carcinogenesis. Morphologic and histochemical changes. *Cancer*. 1982; **50**: 2859–65.
82 Rokitansky A, Trubel W, Buxbaum P, Moeschl P. 1,2-Dimethyl-hydrazine-induced carcinogenesis influenced by different colonic anastomoses in rats. *Eur Surg Res*. 1989; **21**: 184–9.
83 Navarrete A, Spjut H J. Effect of colostomy on experimentally produced neoplasms of the colon of the rat. *Cancer*. 1967; **20**: 1466–72.
84 Wittig G, Wildner G P, Ziebarth D. Der Einfluß der Ingesta auf die Kanzerisierung des Rattendarmes durch Dimethylhydrazin. *Arch Geschwulstforsch*. 1971; **37**: 105–115.
85 Terpstra O T, Dahl E P, Williamson R C, *et al*.. Colostomy closure promotes cell proliferation and dimethylhydrazine-induced carcinogenesis in rat distal colon. *Gastroenterology*. 1981; **81**: 475–80.
86 Byrne P J, Stephens R B, West B, *et al*.. Handling of the bowel during surgery promotes tumour formation in an experimental model. *Br J Surg*. 1988; **75**: 612 (Abstract).
87 Pozharisski K M. The significance of non-specific injury for colon carcinogenesis in rats. *Cancer Res*. 1975; **35**: 3824–30.
88 Calderisi R N, Freeman H J. Differential effects of surgical suture materials in 1,2-dimethylhydrazine-induced rat intestinal neoplasia. *Cancer Res*. 1984; **44**: 2827–30.
89 McCue J L, Phillips R K. Sutureless intestinal anastomoses. *Br J Surg*. 1991; **78**: 1291–6.
90 Risio M, Lipkin M, Candelaresi G, *et al*.. Correlations between rectal mucosa

cell proliferation and the clinical and pathological features of nonfamilial neoplasia of the large intestine. *Cancer Res*. 1991; **51**: 1917–21.

91 Deschner E E, Godbold J, Lynch H T. Rectal epithelial cell proliferation in a group of young adults. Influence of age and genetic risk for colon cancer. *Cancer*. 1988; **61**: 2286–90.

92 Eastwood G L, Trier J S. Epithelial cell renewal in cultured rectal biopsies in ulcerative colitis. *Gastroenterology*. 1973; **64**: 383–90.

93 Serafini E P, Kirk A P, Chambers T J. Rate and pattern of epithelial cell proliferation in ulcerative colitis. *Gut*. 1981; **22**: 648–52.

94 Paganelli G M, Santucci R, Biasco G, *et al.*. Effect of sex and age on rectal cell renewal in humans. *Cancer Lett*. 1990; **53**: 117–21.

95 Risio M, Coverlizza S, Candelaresi G L, *et al.*. Late cytokinetic abnormalities in irradiated rectal mucosa. *Int J Colorectal Dis*. 1990; **5**: 98–102.

96 Deschner E E, Long F C, Hakissian M, Herrmann S L. Differential susceptibility of AKR, C57BL/6J, and CF1 mice to 1,2-dimethylhydrazine-induced colonic tumor formation predicted by proliferative characteristics of colonic epithelial cells. *J Natl Cancer Inst*. 1983; **70**: 279–82.

97 Campbell R L, Singh D V, Nigro N D. Importance of the fecal stream on the induction of colon tumors by azoxymethane in rats. *Cancer Res*. 1975; **35**: 1369–71.

98 Heitman D W, Ord V A, Hunter K E, Cameron I L. Effect of dietary cellulose on cell proliferation and progression of 1,2-dimethylhydrazine-induced colon carcinogenesis in rats. *Cancer Res*. 1989; **49**: 5581–5.

99 Jacobs L R, Lupton J R. Relationship between colonic luminal pH, cell proliferation, and colon carcinogenesis in 1,2-dimethylhydrazine treated rats fed high fiber diets. *Cancer Res*. 1986; **46**: 1727–34.

100 Deschner E E, Cohen B I, Raicht R F. The kinetics of the protective effect of beta-sitosterol against MNU-induced colonic neoplasia. *J Cancer Res Clin Oncol*. 1982; **103**: 49–54.

101 Deschner E E, Wattenberg L W. The proliferative effect of dietary butylated hydroxyanisole on methylazoxymethanol treated colonic mucosa. *Cancer Lett*. 1982; **16**: 197–202.

102 Deschner E E, Alcock N, Okamura T, *et al.*. Tissue concentrations and proliferative effects of massive doses of ascorbic acid in the mouse. *Nutr Cancer*. 1983; **4**: 241–6.

103 Galloway D J. *Dietary manipulation of experimental colorectal cancer*. University of Glasgow: MD Thesis, 1985.

104 Eigler F W, Gross E. Mechanical compression anastomosis (AKA-2) of the colon and rectum. Results of a prospective clinical study. *Chirurg*. 1986; **57**: 230–35.

105 Gross E, Eigler F W. Sutureless compression anastomosis of the distal colon and rectum. An expanded report of experiences with a total of 140 patients. *Chirurg*. 1989; **60**: 589–93.

106 Hardy T G, Jr, Pace W G, Maney J W, *et al.*. A biofragmentable ring for sutureless bowel anastomosis. An experimental study. *Dis Colon Rectum*. 1985; **28**: 484–90.

107 Hardy T G, Jr, Aguilar P S, Stewart W R, *et al.*. Initial clinical experience with a biofragmentable ring for sutureless bowel anastomosis. *Dis Colon Rectum*. 1987; **30**: 55–61.

108 Cahill C J, Betzler M, Gruwez J A, *et al.*. Sutureless large bowel anastomosis: European experience with the biofragmentable anastomosis ring. *Br J Surg*. 1989; **76**: 344–7.

109 Dyess D L, Curreri P W, Ferrara J J. A new technique for sutureless intestinal anastomosis. A prospective, randomized, clinical trial. *Am Surg*. 1990; **56**: 71–5.
110 Stone H H. Nonsuture closure of cutaneous lacerations, skin grafting and bowel anastomosis. *Am Surg*. 1964; **30**: 177–81.
111 Haukipuro K A, Hulkko O A, Alavaikko M J, Laitinen S T. Sutureless colon anastomosis with fibrin glue in the rat. *Dis Colon Rectum*. 1988; **31**: 601–604.
112 Costello A J, Johnson D E, Cromeens D M, *et al.*. Sutureless end-to-end bowel anastomosis using Nd:YAG and water-soluble intraluminal stent. *Lasers Surg Med*. 1990; **10**: 179–84.

9

A comparison between anal endosonography and digital examinations in the evaluation of anal fistulae

Sarah J.D. Burnett and Clive I. Bartram

Introduction

Although many fistulae-in-ano are straightforward to manage surgically,[1] more complex fistula may be complicated by recurrence or postoperative incontinence.[2] A complicated fistula is defined as one where there is more than a simple primary or superficial track. If the full extent of the fistula is not appreciated at surgery, and the fistula is not fully explored, remaining sepsis is bound to cause a recurrence. It might, therefore, be advantageous to determine complex cases preoperatively, so that the case may be allocated more operative time and performed or overseen by an experienced surgeon.

Anal endosonography is a relatively new technique, developed by one of the authors (CIB) as a modification of rectal endosonography. The layers of the anal canal can readily be distinguished, i.e. subepithelial layer (there is no muscularis mucosae and, therefore, the term submucosa is inaccurate), internal sphincter, and external sphincter.[3] It has found an increasing role in the field of coloproctology;[4] one of its earliest applications was in the delineation of the anatomy of fistula-in-ano.[5] Previously, the only radiological investigation was fistulography with water-soluble contrast, which is of little value in routine management.[6] One study reports the use of transrectal endosonography[7] to identify pararectal abscesses.

The pioneer study using anal endosonography was performed at St Mark's hospital and published in 1989.[5] 22 patients were examined blindly after clinical assessment. All seven of the patients found to have complex fistulae were identified; eight out of 12 internal openings were found, and even two foreign bodies were correctly identified! However, none of the extra-, infra- or supralevator tracks were seen. The inability to detect all the internal openings was attributed to the lesions being too close to the probe to be within its effective focal range, and the extrasphincteric tracks being out of range. This original work was highly promising with respect to the endosonographic assessment of fistula, but did not provide information as to how it performed in relation to digital examination, either by the highly trained or the relatively inexperienced surgical finger. To answer this, a prospective trial was instigated.[8]

Methods

Some patients are referred to St Mark's Hospital with perianal sepsis at initial presentation, but many are tertiary referral cases with recurrent sepsis following previous surgery. Over a 4-month period (1.9.89–31.12.89 inclusive) 38 consecutive patients admitted for anal fistulae or perianal sepsis were entered into the study.

Anal endosonography was performed within the 24 hours prior to surgery by one of the two radiologists involved in the trial. The equipment used was a Brüel & Kjær Type 1846 ultrasound scanner that produces a 360 degree radial image, providing a series of axial cuts through the anal canal. The 7 mHz rotating transducer was covered with a hard, sonolucent plastic cone of 1.7 cm external diameter, filled with degassed water for acoustic coupling (Fig. 9.1).

Patients were examined in the left lateral position and the images viewed in the corresponding anatomical plane, so that the 12 o'clock position (i.e., anterior) was on the right of the picture, with the 6 o'clock (i.e., posterior) position on the left (Fig. 9.2). The probe was covered with a condom and ultrasound gel, and gently introduced into the anal canal. A series of images at all levels of the anal canal were then obtained. As the maximum diameter of the plastic cone compares very favourably with that of the average surgeon's finger, patient tolerance for the procedure was good.

The following endosonographic observations were made on each patient:

- The presence of an internal opening and its site
- The presence of an intersphincteric track
- Transsphincteric extension of any intersphincteric track

As previous work has shown that supra- or extrasphincteric tracks, superficial, supralevator and infralevator tracks were not demonstrated

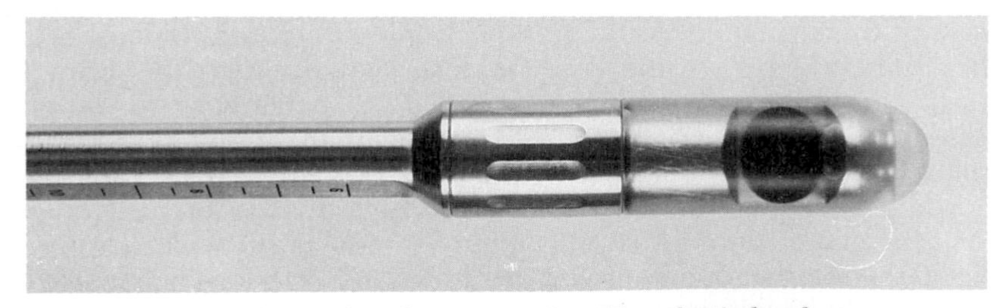

Fig. 9.1 The probe for anal endosonography. The plastic hard cone protects the rotating transducer and is filled with degassed water for acoustic contact.

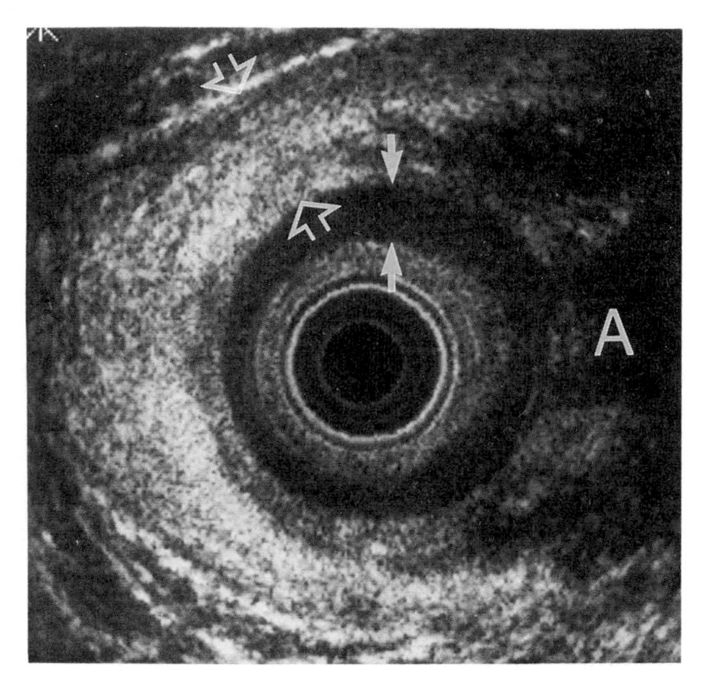

Fig. 9.2 Normal anatomy. The patient is in the left lateral position and the image orientated in the same plane, so that anterior (A) is on the right, and the patient's right side is uppermost. The internal sphincter (small closed arrows) is a clearly defined hypoechoic inner ring, the external sphincter (open arrows) hyperechoic.

with this apparatus, no attempt was made to evaluate these specific aspects.

An early review of the data was conducted as we were unsure of some of the sonographic criteria, particularly for internal openings. Previously, an actual break in the mucosal layer had been used as the sole sign of an internal opening, but on this criteria less than 30 per cent of the internal openings found subsequently at operation were identified.

The review indicated that the following endosonographic features were associated with internal openings:

- A hypoechoic breach of the subepithelial layer of the ano-rectum (the original definition)
- A localized defect in the internal anal sphincter related to an intersphincteric track (in the absence of a previous sphincterotomy)
- The site of contact between an intersphincteric track and the internal sphincter (Fig. 9.3a)

An intersphincteric track was identified as a hypoechoic band running just outside the internal sphincter (Fig. 9.3b). Transsphincteric extension was visualized as a hypoechoic defect within the hyperechoic band of the external sphincter, spreading outside the external sphincter (Fig. 9.4).

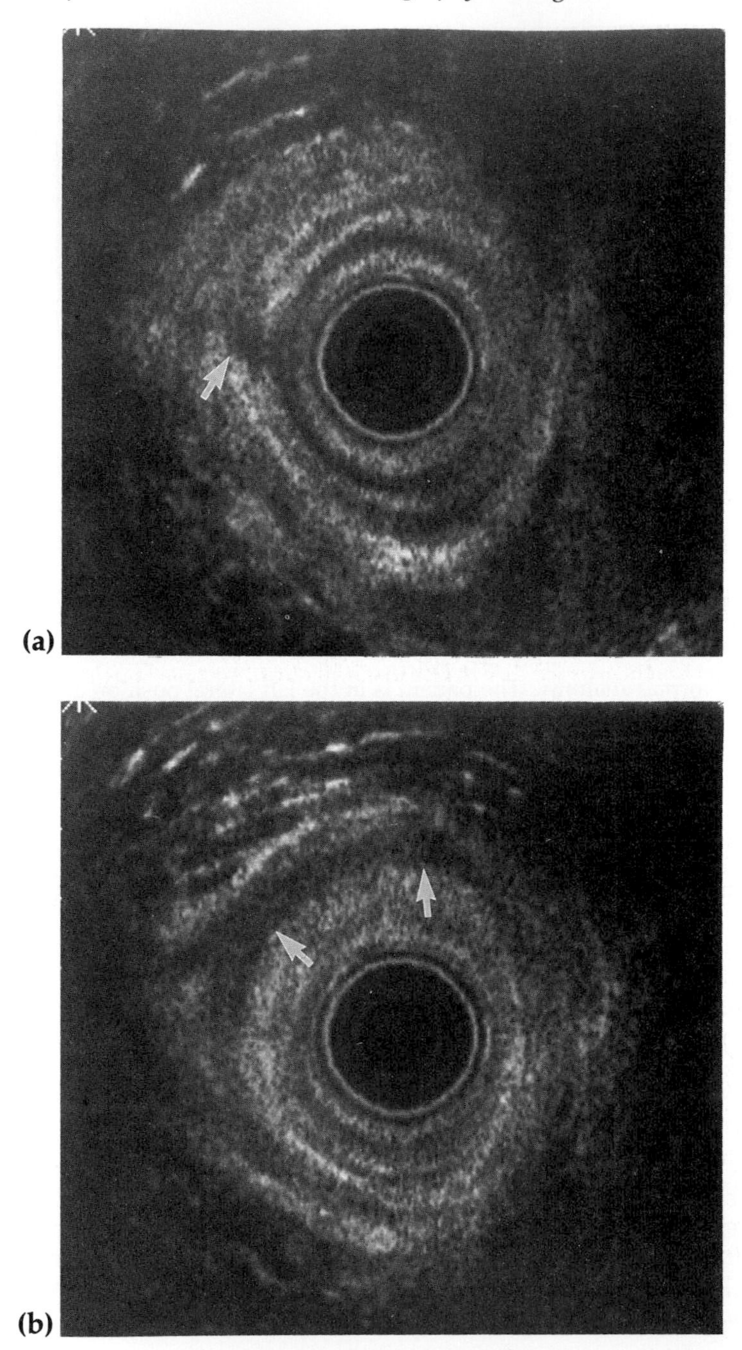

Fig. 9.3 **(a)** Internal opening where the intersphincteric track becomes contiguous with the internal sphincter (arrow).
(b) The intersphincteric track continued higher up as a short horseshoe extension to the right (arrows).

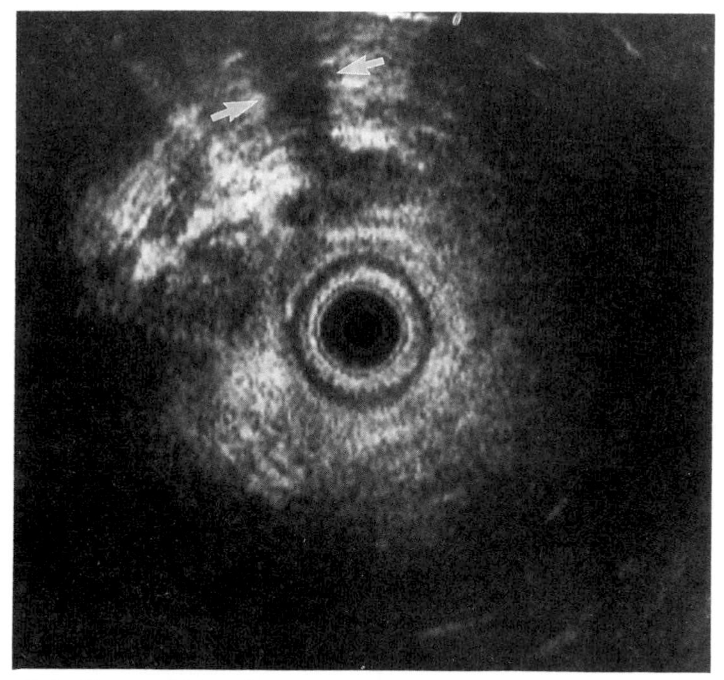

Fig. 9.4 Transsphincteric extension seen as contiguous extension of a hypoechoic track through the external sphincter (arrows).

Differentiation was not made as to whether the tracks were primary or secondary.

The recorded images on the entire series of 38 patients were then reviewed blindly by one radiologist (CIB), using the above criteria, after the clinical component of the study had been completed.

Digital examination was carried out immediately preoperatively with the patient under anaesthetic and in the lithotomy position. Initially the patient was examined by the Surgical Research Fellow, and then by the Consultant in charge of the case. Probes were not used to simulate a routine outpatient clinical examination.

All endosonographic and digital examinations were performed blind, and the information recorded on pro-forma data sheets. Identical sheets were used to record the information from the final review of the ultrasound films. The operative findings were used as the final arbiter of the accuracy of the two assessment techniques. The data were then analysed statistically using the Chi square test with Fisher's exact test.

Results

Anal endosonography was possible in 36 of the 38 patients. Two patients with severe perianal sepsis could not tolerate the procedure because of severe pain.

Internal openings

Surgical exploration revealed a total of 33 internal openings. The Research Fellow and Consultant both correctly identified 26 of the internal openings. Ultrasound identified 10 internal openings using the initial criteria, rising to 24 when the revised criteria were applied. It is noteworthy that in one patient endosonography showed an internal opening not found at operation, but was visible on a fistulogram forwarded from the referring hospital.

Intersphincteric tracks

Of 15 intersphincteric tracks found at operation, ten were correctly identified by anal endosonography, with 11 and ten found by the Surgical Consultant and Research Fellow respectively.

Transsphincteric tracks

14 transsphincteric tracks were found at operation. Of these, 11 were identified at ultrasound, all 14 by the Consultant Surgeon and only 10 by the Surgical Research Fellow.

There was no statistical difference between the clinical and endosonographic assessment in any of these parameters.

Discussion

This study is the first prospective assessment comparing clinical with endosonographic evaluation of anal fistulae, using the operative findings as the 'gold standard'.

As was expected, an experienced colorectal surgeon was able to make an accurate digital assessment of the anatomy of anal fistulae in the majority of cases. The research fellow was able to acquire this skill rapidly during his apprenticeship. The 'database' for clinical examination has been built up over decades of experience and several useful pointers can be learnt. It is important to inspect the perineum very carefully because intersphincteric tracks often open externally very close to the anal verge. Occasionally these may track subcutaneously for a longer distance before discharging. Almost by definition the transsphincteric and other more complicated tracks will open further away, as they have to cross the external sphincter before discharging.

Some aspects of the study were biased. It is undoubtably easier to examine an anaesthetized patient and feel the tracks, than it would be in a routine outpatient. Although the endosonography was performed prospectively, it unfortunately proved essential to review the hard copy of the examinations retrospectively. We are still building up experience in endosonographic interpretation, and felt obliged to undertake this review when the initial criteria for internal openings were found to be so obviously limited. The operative findings were taken as the final arbiter, though these cannot be assumed to be totally correct, and in five of the 38 patients it was impossible to find an internal opening. However, this is obviously an extremely accurate classification of the true extent of the fistula.

Some inherent limitations of ultrasonography and possibly of the equipment became apparent in deciding the limiting factors of endosonographic diagnosis. Lack of mucosal contact with the cone in the ampulla prevented views of suprasphincteric lesions. Changing to the standard rectal balloon system might have overcome this problem. This has been used before,[7] but was not found as helpful as it might have been expected, possibly because the balloon deforms the ampulla too much. The limited penetration of the probe used explains why the full extent of the transsphincteric tracks could not be assessed. New developments in ultrasound technology might permit the focal length to be varied while maintaining a radial image, or differently configured cones could allow visualization of the supralevator compartment. It may also be, however, that with further studies the existing criteria for these lesions will be improved.

Anal endosonography is a new technique that requires considerable experience to interpret often subtle findings. The findings are more complex in patients with previous anal sepsis or surgery. Scars and muscle defects may be seen as hypoechoic bands and wedges, similar to those seen in the active disease process, and on occasion it may be impossible to differentiate the two. In this study, one lesion identified as a transsphincteric track at ultrasound was found surgically to be a scar from the previous operative laying open of a transsphincteric track. Care must also be taken not to

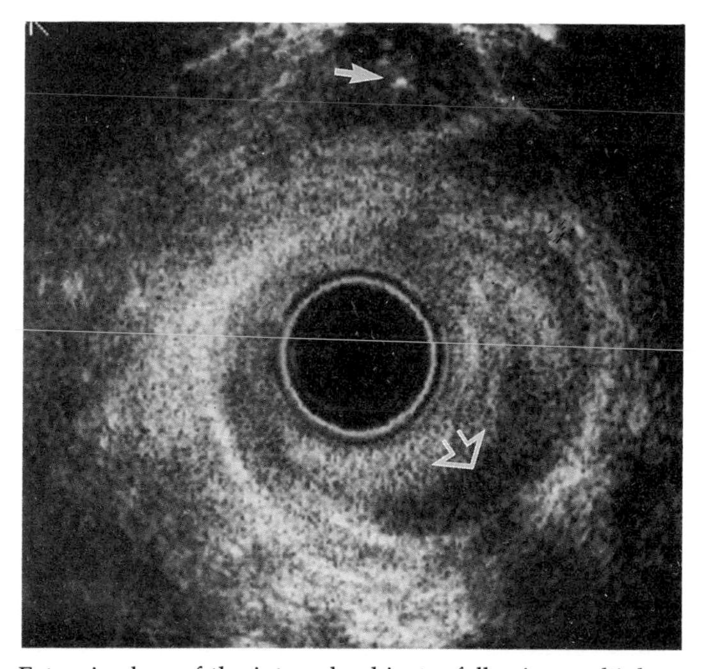

Fig. 9.5 Extensive loss of the internal sphincter following multiple operations for recurrent fistula. Only a few remnants of the internal sphincter (open arrow) are visible and there is gross scarring of the external sphincter. Granulation tissue around a seton (arrow) is noted.

confuse intersphincteric tracks with smooth muscle bands that may run through the longitudinal smooth muscle layer. The experienced operator should be able to identify the muscle bands because they have a more regular appearance, often being bounded by fine hyperechoic lines.

An advantage of endosonographic examination is that the internal sphincter can readily be examined and its integrity assessed. This is of importance because incontinence is always a risk with recurrent fistula surgery (Fig. 9.5). Internal sphincter defects were found in a large number of our patients, both consequent on previous surgery and in those with active sepsis, which seems to destroy the smooth muscle of the internal sphincter. The presence of induration in the anal canal may render digital assessment of the muscle bulk of the internal sphincter difficult. Awareness of the pre-operative condition of the internal anal sphincter may alert the surgeon to attempt to preserve it.

Expert clinical examination has been confirmed as a valid preoperative means of assessing complex fistula. The role of anal endosonography is yet to be fully established. Technical developments may extend its field of view and overall accuracy. At present an important role is to assess the integrity of the sphincters, especially the internal sphincter, in patients with recurrent fistula problems.

Acknowledgements

The authors would like to thank Mr R J Nicholls, Consultant Surgeon at St Mark's Hospital and Mr Seow Choen, Research Fellow for organizing this study and undertaking its clinical components.

References

1 Shouler R J, Grimley R P, Keighley M R B, Alexander-Williams J. Fistula-in-ano is usually simple to manage surgically. *Int J Colorect Dis*. 1986; **1**: 113–5.

2 Goligher J C (Moderator). Symposium: Fistula-in-ano. *Int J Colorect Dis*. 1987; **2**: 51–71.

3 Law P J, Bartram C I. Anal Endosonography: technique and normal anatomy. *Gastrointest Radiol*. 1; 1989: 349–53.

4 Bartram C I, Burnett S J D. *Atlas of Anal Endosonography*. London: Butterworth Heinemanns: 1991.

5 Law P J, Talbot R W, Bartram C I, Northover J M A. Anal endosonography in the evaluation of perianal sepsis and fistula-in-ano. *Br J Surg*. 1989; **76**: 7152–755.

6 Kuijpers H C, Schulpen T. Fistulography for fistula-in-ano. Is it useful? *Dis Col Rect*. 1985; **28**: 103–104.

7 Cammarota T, Discalzo L, Corno F, *et al.*. First experiences with transrectal echotomography in perianal abscess pathology. *Radiol Med*. (Torino) 1986; **72**: 837–40.

8 Choen S, Burnett S, Bartram C I, Nicholls. Comparison between anal endosonography and digital examination in the evaluation of anal fistulae. *Br J Surg*. 1991; **78**, 445–7.

10

The postoperative recurrence of Crohn's disease

Adam D.N. Scott and Robin K.S. Phillips

Introduction

Despite considerable research interest and activity, the aetiology and pathogenesis of Crohn's disease remain unclear. It has not, therefore, been possible to develop effective medical treatment for the disease, hence up to 95 per cent of patients require surgical treatment at some time.[1] Unfortunately the risk of recurrence after surgery is very high – approximately 60 per cent at 10 years[2] – and it is important that surgeons should be aware of any factors that might influence this risk of recurrence. This chapter reviews the importance of some of the reported factors.

In addition, the records of the 241 patients who were treated by right hemicolectomy for ileal or ileocolic Crohn's disease at St Mark's between 1947 and 1988 have been reviewed. During this time a total of 1609 patients with Crohn's disease, who had not undergone intestinal resection, were referred to St Mark's. None of the 241 patients had had a previous resection for Crohn's disease but 44 patients were excluded from further analysis because of residual disease identified by the surgeon (20), inadequate pathological or follow-up data (22) or perioperative death (2). The results of surgery in the remaining 197 patients will be used to illustrate the significance of blood transfusion, granulomas, resection margin involvement and extent of disease in determining the risk of recurrence. The importance of an anastomosis will also be discussed. To start with, there must be an understanding of what is meant by recurrent disease.

The definition of recurrence

This problem was first highlighted by Lennard-Jones and Stalder[3] who pointed out that at least three clinical situations could be called 'recurrence':

- A recurrence of clinical symptoms in a patient with no other signs of new disease
- A recurrence of symptoms in a patient with radiological and/or histological evidence of recurrence, but not requiring further surgery

- Recurrent disease which required further surgery

They clearly showed that the rate of recurrence diminished as more stringent definitions of recurrence were used. However, this system of definition still does not distinguish between new disease in a previously normal area of bowel, and persistence or recrudescence of disease that had been left *in situ* at the time of the original operation.[4,5] The situation is further complicated by the fact that spontaneous resolution of Crohn's disease can occur and the disease can sometimes improve or even disappear with medical treatment.

It has also been shown that after faecal diversion for colonic disease granulomas may still be found in the colon although other signs of disease such as mucosal inflammation and ulceration have disappeared.[6] Does this then represent healing of disease? If it does, a further flare-up may be called a recurrence. If, however, it means that disease is still present but quiescent, then a further flare-up may be called a relapse.

Such points cannot really be answered until the underlying cause of the disease is known. For instance, it has been suggested that the granuloma represents a host response to the prime aetiological agent, and that the development of fissures and fistulae are due to super-added infection with anaerobic organisms. If this is the case, then the persistence of a granuloma may not mean that the disease is still present. On the other hand, if the granuloma is only present because the macrophages within it are unable to digest some particulate material,[7] then persistence of the granuloma implies persistence of the disease, even though it may be quiescent at the time.

As regards the postoperative recurrence of Crohn's disease, it has, for a long time, been appreciated that some symptoms that may otherwise suggest recurrent disease may be the result of the operation itself.[5] Right hemicolectomy is often followed by a change in bowel habit; after extensive resection a simple gastroenteritis may result in illness suggesting recurrent disease. A definition of recurrence that relies only on clinical symptoms is therefore likely to overestimate the true incidence of recurrence. Conversely, a definition that relies on surgical excision for proof will underestimate the recurrence rate because many patients with proven recurrent disease will not require further resection.

These various definitions of recurrence all include clinical symptoms and therefore ignore patients with asymptomatic recurrent disease. It does not seem reasonable in routine clinical practice to screen asymptomatic patients for evidence of recurrent disease because no evidence that treatment of such disease is beneficial exists. However, endoscopy has been used to study the natural history of recurrence in 114 patients who had undergone 'curative' resection of ileocolic disease.[8] This demonstrated that the true rate of recurrence postoperatively was 72 per cent at one year – much higher than previously imagined – and 42 per cent of the patients with endoscopic evidence of recurrent disease were asymptomatic. This study provides a fourth definition of recurrence which is more accurate (but less practical) than any of the other three:

- Patients with or without symptoms who have endoscopic and histological evidence of Crohn's disease after previous 'curative' surgery

For the purposes of our analysis we have used the second definition of recurrence i.e., a recurrence of symptoms in a patient with radiological and/or histological evidence of recurrent disease but not necessarily requiring further surgery.

The method used to calculate the recurrence rate is also important. Crude recurrence rates ignore the length of follow-up and, therefore, underestimate the risk of recurrent disease[3] because the rate of recurrence will appear low in patients followed for only a few years. Life table or actuarial analysis[9,10] takes variations in follow-up into account and allows a cumulative recurrence rate to be calculated for each year of follow-up based on the number of patients at risk of recurrence during that year.

Age and sex

It used to be felt that the risk of recurrence was related to the age of the patient at onset of the disease. One study of 52 patients showed a recurrence rate of 62 per cent in those under 30 years of age and 30 per cent in those over 30 years,[11] and a British study involving 172 patients demonstrated a greatly increased risk of recurrence in those less than 15 years of age.[4] However, most actuarial studies have shown no correlation between age of onset and subsequent recurrence rates.[12,13] In Birmingham the re-operation rate in 67 patients younger than 16 years at onset of disease was similar to that in patients who developed the disease in adulthood[14], and Greenstein *et al.*[15] showed that although the recurrence rate 5 years postoperatively seemed to be higher in younger patients (less than 25 years at onset of disease) this difference disappeared with longer follow-up.

It has also been suggested that the age of the patient at operation is related to the risk of recurrence. A Swedish study[12] demonstrated a significantly higher recurrence rate at 10 years in patients less than 25 years old at operation (55 per cent) when compared with patients over 40 years old at the time of operation (40 per cent). However, most studies have not shown any correlation between risk of recurrence and age at operation[2,3,16] and in our own study at St Mark's there was no significant relation between the age of the patient at the time of resection and the incidence of recurrence.

In Europe and the UK Crohn's disease is more common in women than men.[17] Although some authors have reported a worse prognosis in women,[5,18] most studies, including our own series at St Mark's, show no difference in the recurrence rate.[2,11,16]

Duration of disease

It has been claimed that a short preoperative history is associated with an increased risk of postoperative recurrence[3,5,13] and that this implies that a short history and early recurrence are both manifestations of a particularly aggressive form of the disease.[5] However, the opposite effect has also been observed with a higher recurrence rate in those with a longer history[2,19] and others have shown no association between crude recurrence rates and length of preoperative history.[12]

Although the interval between diagnosis and operation can be reliably ascertained even in retrospective analyses, the time between onset of disease and diagnosis is notoriously difficult to measure because the symptoms of Crohn's disease are so variable in severity and it is generally accepted that many patients with proven disease are completely asymptomatic.[8] In view of this difficulty and the lack of consistency in the series quoted above it seems unlikely that there is any significant association between length of history and postoperative recurrence.

Resection margin involvement

The biphasic recurrence curve demonstrated by the Leeds group[5] had suggested that early recurrences may be due to the rapid growth of microscopic lesions at the resection margins of the retained bowel. This prompted a number of groups to explore the role of resection margin frozen section as advocated by Kyle[20] in the surgical management of Crohn's disease.[21,22] This approach involved frozen section examination of resection margins with further resection of additional lengths of intestine until histologically normal margins were found. The rationale behind this approach depended on two major assumptions: firstly, that frozen section examination is a reliable method of diagnosing resection margin involvement and, secondly, that the patient's clinical outcome is improved by the selection of histologically normal resection margins. However, both these assumptions are probably unjustified.

Frozen section examination appears to be a poor method of assessing resection margin involvement. When 60 resection margins, which had been reported as normal after frozen section examination, were re-examined together with paraffin sections from the resected specimen, 19 were found to be involved due to Crohn's disease.[22] This was largely due to sampling error because the inflammation in Crohn's disease is focal and discrete[23] and can therefore be easily missed. Indeed, in 13 of the misdiagnosed resection margins, the frozen section slides simply did not show the margin involvement that was apparent when all available histological sections of the margin were reviewed. However, this was not the only problem, because in the remaining six misdiagnosed frozen section slides retrospective review demonstrated active inflammation and granulomas that had not initially been recognized.

In any case, it has been claimed that inflammation at the resection margins does not influence subsequent recurrence.[2,24] This was first demonstrated by the Oxford group[21] who showed that active inflammation at the resection margins had no influence on either the incidence or timing of recurrent disease – active inflammation in their study meaning a characteristic histological picture rather than non-specific changes such as minor fibrosis, oedema or inflammation, which could be the result of intestinal obstruction. This view has been supported by some studies[2,24,25] and contradicted by others.[26] Softley[27] believes that, in patients with a clearance of 10 cm or more between the excisional margin and diseased tissue, recurrence is three times less likely than in patients with a smaller margin of clearance.

The results of detailed histological examination of the resection margins of the right hemicolectomy specimens were available in 151 of our study group of 197 patients. The results are shown in Table 10.1 and demonstrate that the recurrence rate is significantly higher when there is macroscopic disease at either resection margin. These results would support the view that end-to-end anastomoses, which are constructed in the macroscopically abnormal bowel, are very likely to give rise to early recurrence. This is an important observation, as surgeons who use strictureplasty (and who have been impressed with its results) may adopt a more cavalier attitude towards the macroscopically abnormal bowel than this evidence permits.

The situation as regards microscopic disease is unclear. The recurrence rate of 40 per cent in patients with microscopically involved resection margins is not significantly different from the rate of 56 per cent in those with macroscopic disease or that of 31 per cent in the patients without resection margin involvement. However, the number of patients with either macroscopic ($n = 16$) or microscopic ($n = 20$) disease at the resection margins is small and may be insufficient to demonstrate statistical significance, even if the difference is real. Whether or not microscopic disease at the resection margin has a significant effect on the incidence of recurrence, the inaccuracy of frozen section examination in this situation means that microscopic disease at the resection margin cannot be used to guide the surgeon. However, postoperatively it may help to identify those patients at greatest risk of developing recurrent disease.

It is difficult to reconcile these findings in our own patients with the success of strictureplasty, which was first described by Emmanoel Lee in Oxford at a time when active disease at the suture line was not thought to influence subsequent recurrence.[28] It is now well established that strictureplasty in the presence of disease is associated with very few leaks[29,30] and a low recurrence rate.[31] It is possible that the tight strictures, which have been selected for treatment by strictureplasty, represent primarily fibrotic disease and it has been suggested that strictureplasty in areas of active disease may be followed by a high recurrence rate.[32]

Table 10.1 Resection margin involvement and recurrence

Resection margins	Recurrence (n)	No recurrence (n)	Recurrence rate (%)
Macroscopic disease	9	7	56*
Microscopic disease	8	12	40
Normal	36	79	31*

*Significantly different ($X^2 = 3.87$, $p < 0.05$).

Extent of disease

A number of early studies showed that the recurrence rate was higher in patients with extensive disease of the bowel.[33] The Mayo Clinic reported 285 patients followed up for more than 2 years. The recurrence rate was approximately 60 per cent until the inflammatory lesion reached 50 cm in length, but with lesions over 50 cm the recurrence rate was 76 per cent and, with lesions over 80 cm, the rate of recurrence was 93 per cent.[19] The first major British study[4] from Leeds also showed a clear relation between the rate of recurrence and the length of the inflammatory lesion. With lesions less than 7.5 cm the recurrence rate was 38 per cent but rose to 67 per cent if more than 30 cm of bowel were affected. Other authors have also reported an association between the length of diseased bowel and the risk of recurrence,[11,34] although Trnka *et al.*[35] compared 23 patients who developed recurrent disease with 13 who did not and found no difference in the extent of inflammation.

We have reviewed the histological records of the patients who underwent right hemicolectomy for ileal or ileocolic Crohn's disease at St Mark's between 1947 and 1988. The association between the length of involved bowel as measured by the pathologist from the resected specimen and the risk of developing recurrent disease is shown in Fig. 10.1. The recurrence rates quoted are crude rates but there is no difference in follow-up between the groups of patients. It is clear that there is a strong positive correlation between extent of disease and risk of recurrence.

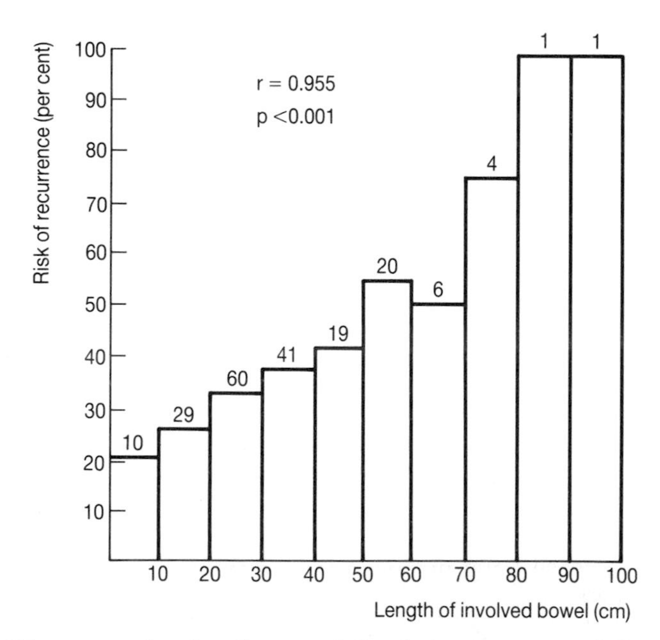

Fig. 10.1 Histogram showing the association between extent of disease and risk of recurrence (r = correlation coefficient). (The number at the top of each column represents the number of patients in that group.)

Perioperative blood transfusion

It has been suggested that perioperative blood transfusion might have some influence on the development of recurrent disease. This suggestion is based on the observation that blood transfusion seems to have a significant effect in at least two different clinical situations: renal transplantation and cancer surgery.

It is nearly 20 years since the beneficial effect of blood transfusion in prolonging renal allograft survival was first demonstrated.[36] Although the mechanism of this 'transfusion effect' is still not clear,[37] it is likely to involve not only a non-specific down regulation of the immune system,[38,39] but also specific immunological effects. These might include the deletion of particular clones of lymphocytes,[40] the stimulation of suppressor T-cells[41,42] and the appearance of anti-idiotypic[43] and blocking[44] antibodies, which, by binding to the antigen-combining site of an antibody, prevent that antibody from interacting with its particular antigen.

More recently it has been suggested that the immunomodulatory effect of blood transfusion might increase the risk of local recurrence[45–48] and infectious complications[49] after surgery for colorectal cancer. The evidence for such an effect on other tumours is less clear,[50–55] but there is certainly a greatly increased incidence of malignancy when the immune system is compromised in either immunodeficient[56,57] or immunosuppressed[58,59] patients.

The effects of blood transfusion in Crohn's disease have been reported in two studies, both of which have suggested that blood transfusion protects against the development of recurrent disease. The first study[60] was a 4-year follow-up of 79 patients, of whom 49 per cent had ileocolic disease and 27 per cent colonic disease only. Transfusion at the time of surgery produced a significantly lower recurrence rate both in the group as a whole (20 versus 44 per cent) and in the subgroup with ileocolic disease (5 versus 45 per cent). However, in a subsequent report, statistical significance for the ileocolic group had been lost.[61] The second study involved 60 patients with small bowel disease followed up for 5 years and this also showed a significantly lower recurrence rate in the transfused patients (19 versus 59 per cent).[62]

Of the 197 patients undergoing right hemicolectomy for Crohn's disease at St Mark's between 1947 and 1988, 102 (52 per cent) received a blood transfusion perioperatively. The transfused and non-transfused patients were broadly comparable although there were significant differences in the preoperative haemoglobin and albumin concentrations, as might be expected. Fig. 10.2 shows the cumulative rates of recurrence in the two groups of patients and demonstrates no significant difference (logrank test).

The immunological effects of blood transfusion in patients with inflammatory bowel disease have not been investigated. Certainly there is evidence that nutritional disturbances,[63] treatment with steroids[64] and Crohn's disease itself[65] may all be associated with reduced immunological responsiveness, and all these factors may be relevant in patients requiring surgery for complicated Crohn's disease.

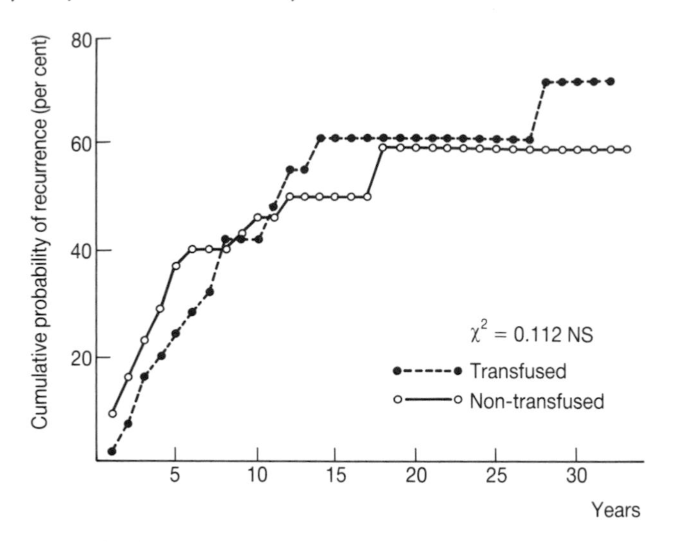

Fig. 10.2 Graph showing the cumulative recurrence rate in transfused and non-transfused patients. Reproduced by permission of the publishers, Butterworth-Heinemann Ltd from Scott *et al.*, *Br J Surg*. 1991; **78**: 455–8.

The timing of transfusion may be of critical importance – experimental studies have shown an accelerated rate of tumour growth in animals when allogeneic blood is given before transplantation of a tumour, but no effect is observed if the blood is given at the time of, or after, tumour transplantation.[66,67] Potential renal transplant recipients are transfused before receiving the antigenic stimulus of the allograft and long-term immunosuppression, whereas transfused cancer patients receive blood at the time of surgery, presumably many months after the development of a tumour, which may already be associated with immunological impairment.[68] It is therefore likely that the effects of transfusion in transplantation and cancer involve different immunological mechanisms none of which may be relevant in inflammatory bowel disease.

It is possible that the immunological effects of Crohn's disease alter the usual response of the immune system to blood transfusion and thereby prevent it from having any clinical effect. If this is the case, then it may be that transfusion might have a more pronounced effect in patients with mild disease and less of an immunological disturbance. After surgery for Crohn's disease the risk of recurrence is extremely high – more than 60 per cent at 10 years[2] – and the question of a transfusion effect in surgery for Crohn's disease therefore deserves further investigation. Ethical and practical considerations preclude a prospective randomized clinical trial but it would be possible to investigate the effect of transfusion, and the timing of its administration relative to surgery, on aspects of immune function in Crohn's disease.

Granulomas

The clinical significance of the granuloma in Crohn's disease is unclear. It has been suggested that an absence of granulomas in the resection specimen after 'curative' surgery for Crohn's disease favours the subsequent development of recurrent disease. In a study of 154 patients, patients without evidence of granuloma formation were twice as likely to suffer a recurrence as those in whom granulomas were found.[27] Chambers and Morson[69] found in a group of 79 patients over a 13-year period that a high content of granulomas predicted a good prognosis in large bowel Crohn's disease, but was of no prognostic significance in small bowel disease. They also noted a wide variation in the granuloma counts – one per section in the small bowel, six in the colon, 18 in the rectum and 36 in the anal canal – and noted that a long clinical history was associated with a low granuloma content. Their findings were consistent with the view that the granuloma represents an adaptive mechanism for the removal or localization of the causative agent of Crohn's disease.

Similar results were reported in a study of 90 patients, 38 of whom required re-operation for recurrent disease. When the patients were divided into those with granulomas and those without, it was found that the recurrence rate was 34 per cent in those with granulomas and 60 per cent in those without ($p<0.025$). Whether the granulomas were found in the bowel wall or in the mesenteric lymph nodes appeared to make no difference.[70]

However, in another study of 102 patients, it was reported that there was no difference in recurrence-free survival between the 53 patients with granulomas and the 49 without.[71] These authors also noted that the likelihood of finding granulomas was influenced by the site of disease – granulomas were found in 45 per cent of patients with ileitis, 53 per cent with colitis and 58 per cent with ileocolitis. Others have also found that granulomas, whereas they tend to occur in younger patients, do not appear to protect against recurrent disease,[72] so that the clinical importance of granulomas remains unclear.

The histological records of the 197 patients who had undergone right hemicolectomy for Crohn's disease at St Mark's between 1947 and 1988 were reviewed. Granulomas were found in the bowel wall of 87 patients (44 per cent) and in the mesenteric nodes of 40 patients (20 per cent) and the recurrence rates in these patients are shown in Table 10.2.

These results certainly do not support the view that granulomas protect against recurrent disease. In fact, the recurrence rates in patients with granulomas appear to be higher than the rates in patients without evidence of granuloma formation, although these differences do not reach statistical significance. However, it is interesting that patients with granulomas in either the bowel wall or nodes were significantly younger than patients without granulomas. It does appear that the immunohistochemical character and pattern of intestinal macrophages may change with increasing age and it is possible that their functional activity is also altered.

Chemotactic and random migration of intestinal macrophages obtained by enzymatic digestion of resection specimens have been tested in a

Table 10.2 Recurrence rates and granulomas

	Recurrence (n)	No recurrence (n)	Total patients	Recurrence rate (%)
Bowel wall				
granuloma	35	52	87	40
no granuloma	30	67	97	31
Lymph nodes				
granuloma	18	22	40	45
no granuloma	41	75	116	35

Boyden chamber. This work demonstrated that, in patients with Crohn's disease, the motility of intestinal macrophages was significantly lower in patients without granulomas than in those with granulomas.[73] It is therefore possible that the formation of granulomas in Crohn's disease is related to the ability of the macrophages to accumulate at the site of inflammation.

In general, it would appear that the macrophages of younger patients are better able to move to the inflammatory focus than are the macrophages of older patients. However, this does not seem to produce any clinical advantage for the younger patient (see above). This implies that the intestinal macrophages, although they are able to accumulate at the site of inflammation, are unable to deal with the underlying cause of the inflammation and lends weight to the suggestion that the intestinal macrophages in Crohn's disease are in some way abnormal.[74]

The proportion of our patients with granulomas in the bowel wall was significantly higher in those with ileocolic disease (57 per cent) than in those with ileal disease only (40 per cent). The association between the site of disease and presence of granulomas in the bowel wall is interesting and is consistent with previous studies. Chambers and Morson[69] noted that the number of granulomas seen in a series of patients with Crohn's disease affecting either the terminal ileum or colon was always greater in the latter, perhaps reflecting immunological differences between large and small bowel.

Anastomosis

Patients with colonic Crohn's disease are treated either by proctocolectomy and ileostomy or colectomy and ileorectal anastomosis. In some series these two groups are not directly comparable because patients undergoing proctocolectomy may be more likely to have perianal and severe rectal disease than those selected for ileorectal anastomosis.[75] However, Fig. 10.3 shows the cumulative recurrence rates after ileorectal anastomosis or proctocolectomy for Crohn's colitis at St Mark's, and patients with severe rectal or ileal disease have been excluded. The difference in recurrence rates between these two groups is highly significant and indicates that a

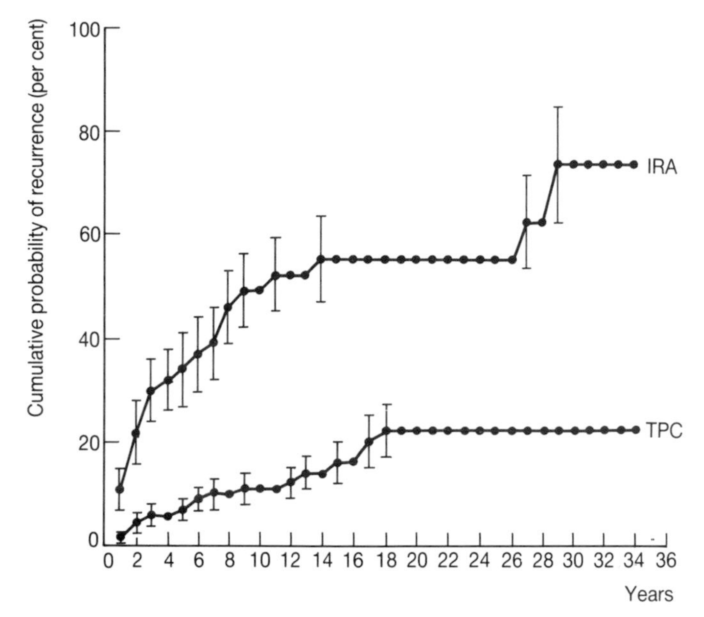

Fig. 10.3 Graph showing the cumulative recurrence rate after total proctocolectomy (TPC) or colectomy and ileorectal anastomosis (IRA) for Crohn's disease at St Mark's Hospital 1947–87. Reproduced by permission from Ritchie, *Ann R Coll Surg*. 1990 **72**: 155–7.

major factor in determining the risk of recurrence after surgery for Crohn's disease is the presence or absence of an anastomosis. This is entirely consistent with the observation that recurrent disease, when it does occur, is almost invariably found in the region of the anastomosis.[5,8] What is it about an anastomosis that makes it such a favoured site for recurrent disease?

It has been suggested that the mechanism of recurrence at an anastomosis might involve free reflux of a large bowel or faecal factor[76] across the anastomosis into the small bowel and in the same way small bowel content might damage the rectum.[77] However, in 11 patients treated by left hemicolectomy or sigmoid colectomy with an anastomosis between two segments of large bowel, the recurrence rate was as high as that following an ileorectal anastomosis.[78] This is also true of small bowel anastomoses after resection of jejunal or ileal disease.[1,79] This evidence would not support the view that free reflux of large bowel content into the small bowel, or vice versa, is responsible for anastomotic recurrence.

There may be some delay in intestinal transit across an anastomosis, particularly when there is a degree of stenosis at the suture line. This is akin to the physiological 'ileal brake' that occurs at the ileocaecal valve and results in more prolonged exposure of the mucosa of the terminal ileum to intestinal content. Stasis might also occur on the proximal side of an anastomosis and, by the same mechanism, lead to recurrent disease. If this is the case, then recurrence should be restricted to the proximal side of the anastomosis. In a study of 83 patients followed for 5 years, the recurrence rate after resection was 26.5 per cent and these recurrences were indeed found proximal to the anastomosis.[80] Similar results were reported in a study of 20 patients with recurrent disease, in 18 of whom the recurrence was proximal to the anastomosis.[81] Farmer *et al.*[82] also found that, in a group of 615 patients, the recurrences were proximal to the anastomosis, whether this be ileorectal or ileocolic.

On the other hand, Flint *et al.*[83] reported that, among 12 patients with recurrent disease, the recurrence was most commonly found on both sides of the anastomosis. Koch *et al.*,[84] in a study of 56 patients after ileocolic resection and anastomosis, found that the site of recurrence was related to the site of the original disease; those with terminal ileal disease developed recurrences proximal to the anastomosis, whereas recurrences in those with ileocolic disease were found on both sides of the anastomosis. It would therefore appear that recurrence is not necessarily restricted to the proximal side of the anastomosis.

It is possible that the defect in the mucosa following division of the bowel and anastomosis allows luminal contents and antigens to come into direct contact with the mucosal immune system of the gastrointestinal tract in the lamina propria. This could produce inflammation, which in turn might produce the characteristic features of Crohn's disease. However, the mucosal defect probably only exists for a short period of time after anastomosis and yet recurrent disease can develop many years later. In addition, the combination of bowel preparation in the case of large bowel disease, preoperative fasting and postoperative ileus, will usually keep the bowel empty for the short time after anastomosis during which a mucosal defect might exist.

A study of mesenteric vasculature in 15 patients with Crohn's disease, using a technique of vascular casting, has suggested that ischaemia may be involved in the pathogenesis of Crohn's disease. Examination of these specimens suggested a sequence of events, beginning with vascular injury and ending with multifocal tissue infarction.[85] The blood supply of experimental anastomoses has been studied using techniques of silicone rubber casting,[86] measurements of oxygen tension,[87] temperature measurement and implanted ultrasonic Doppler flowmeters.[88] There is little doubt that intestinal anastomoses may be associated with ischaemia and this could affect both sides of the anastomoses. However, it is again difficult to relate ischaemia at the time of constructing an anastomosis to the development of recurrent disease, often many years later.

Suture materials are present on both sides of an anastomosis, may be present for many years, are associated with a variable inflammatory response, pass through the intestinal wall leaving a track from the lumen into the lamina propria and submucosa, and in some situations are known to give rise to granulomas,[89] possibly indicating a local effect on immunological mechanisms. Experimental work in the field of colorectal cancer has clearly shown that suture materials can exert a long-term biological effect *in vivo* and influence the development of carcinogen-induced tumours.[90,91] It is therefore possible that suture materials play a role in the development of perianastomotic disease. Although our study of patients undergoing right hemicolectomy at St Mark's did not demonstrate any association between the type of material used and the subsequent recurrence rate, experimental work does suggest that all currently available materials affect local immunological mechanisms. In the future it may be possible to develop immunologically-inert suture materials – certainly this avenue of research is worth exploring because the problem of anastomotic recurrence is one of the burning issues in the modern management of Crohn's disease.

Can the surgeon influence the risk of recurrence?

Many of the factors discussed above are outside the control of the surgeon but it is possible to draw some conclusions that are of importance to the surgeon operating for Crohn's disease.

- Every effort should be made to avoid constructing the anastomosis in an area of macroscopic disease. The presence of microscopic disease is undesirable but may be inevitable and frozen section examination of the resection margins is not worthwhile
- When operating for colonic disease an ileorectal anastomosis (when feasible) avoids the potential morbidity of pelvic dissection, a perineal wound and a stoma but at the expense of a significantly higher recurrence rate and the patient should be made aware of this
- There would appear to be no place for routine transfusion or early operation in the absence of complications because no significant association has been demonstrated between recurrence rate and either transfusion status or length of preoperative history

References

1. Bergman L, Krause U. Crohn's disease – a long term study of the clinical course in 186 patients. *Scand J Gastroenterol.* 1977; **12**: 937–44.
2. Speranza V, Simi M, Leardi S, Del Papa M. Recurrence of Crohn's disease after resection. Are there any risk factors? *J Clin Gastroenterol.* 1986; **8**: 640–46.
3. Lennard-Jones J E, Stalder G A. Prognosis after resection of chronic regional ileitis. *Gut.* 1967; **8**: 332–6.
4. Atwell J D, Duthie H L, Goligher J C. The outcome of Crohn's disease. *Br J Surg.* 1965; **52**: 966–72.
5. De Dombal F T, Burton I, Goligher J C. Recurrence of Crohn's disease after primary excisional surgery. *Gut.* 1971; **12**: 519–27.
6. Thompson H. Rectal biopsies from the bypassed rectum. *Arch Fr Malad Appareil Dig Nutr.* 1974; **63**: 583–5.
7. Williams D, Williams W J, Williams J E. Enzyme histochemistry of epithelioid cells in sarcoidosis and sarcoid-like granulomas. *J Pathol.* 1969; **97**: 705–709.
8. Rutgeerts P, Geboes K, Vantrappen G, *et al.*. Natural history of recurrent Crohn's disease at the ileocolonic anastomosis after curative surgery. *Gut.* 1984; **25**: 665–72.
9. Hill A B. *Principles of medical statistics* (8th edn). London: The Lancet, 1966, pp. 232–6.
10. Armitage P. *Statistical methods in medical research.* Oxford: Blackwell Scientific Publications, 1974, pp. 408–414.

11 Stahlgren L H, Ferguson L K. The results of surgical treatment of chronic regional enteritis. *J Am Med Assoc*. 1961; **175**: 986–9.
12 Hellers G. Crohn's disease in Stockholm county 1955–74. A study of epidemiology. Results of surgical treatment and long-term prognosis. *Acta Chir Scand* [Suppl]. 1979; **490**: 31–48.
13 Higgens C S, Allan R N. Crohn's disease of the distal ileum. *Gut*. 1980; **21**: 933–40.
14 Puntis J, McNeish A S, Allan R N. Long-term prognosis of Crohn's disease with onset in childhood and adolescence. *Gut*. 1984; **25**: 329–36.
15 Greenstein A J, Sachar D B, Pasternack B S, Janowitz H D. Reoperation and recurrence in Crohn's colitis and ileocolitis. Crude and cumulative rates. *N Engl J Med*. 1975; **293**: 685–90.
16 Nygaard K, Fausa O. Crohn's disease: recurrence after surgical treatment. *Scand J Gastroenterol*. 1977; **12**: 577–84.
17 Devlin H B, Datta D, Dellipiani A W. The incidence and prevalence of inflammatory bowel disease in North Tees Health District. *World J Surg*. 1980; **4**: 183–93.
18 Truelove S C, Pena A S. Course and prognosis of Crohn's disease. *Gut*. 1976; **17**: 192–201.
19 Van Patter W N, Bargen J A, Dockerty M B, *et al*.. Regional enteritis. *Gastroenterology*. 1954; **26**: 347–450.
20 Kyle J. Surgical treatment of Crohn's disease of the small intestine. *Br J Surg*. 1972; **59**: 821–3.
21 Lee E C G, Papaioannou N. Recurrences following surgery for Crohn's disease. *Clin Gastroenterol*. 1980; **9**: 419–38.
22 Hamilton S R, Reese J, Pennington L, *et al*.. The role of resection margin frozen section in the surgical management of Crohn's disease. *Surg Gynecol Obstet*. 1985; **160**: 57–62.
23 Yardley J H, Hamilton S R. Focal non-specific inflammation in Crohn's disease. In Pena A S, Weterman I T, Booth C C, Strober W. (Eds), *Recent advances in Crohn's disease. Developments in Gastroenterology* (vol. 1) The Hague: Martinus Nijhoff, 1981, pp. 62–6.
24 Heuman R, Boeryd B, Bolin T, Sjodahl R. The influence of disease at the margin of resection on the outcome of Crohn's disease. *Br J Surg*. 1983; **70**: 519–21.
25 Pennington L, Hamilton S R, Bayless T M, Cameron J L. Surgical management of Crohn's disease. Influence of disease at margin of resection. *Ann Surg*. 1980; **192**: 311–8.
26 Wolff B G, Beart R W, Frydenberg H B, *et al*.. The importance of disease-free margins in resections for Crohn's disease. *Dis Colon Rectum*. 1983; **26**: 239–43.
27 Softley A, Myren J, Clamp S E, *et al*.. Factors affecting recurrence after surgery for Crohn's disease. *Scand J Gastroenterol*. 1988; **23** (Suppl 144): 31–4.
28 Lee E C G, Papaioannou N. Minimal surgery for chronic obstruction in patients with extensive or universal Crohn's disease. *Ann R Coll Surg Engl*. 1982; **64**: 229–33.
29 Dehn T C B, Kettlewell M G W, Mortensen N J McC, *et al*.. Ten-year experience of strictureplasty for obstructive Crohn's disease. *Br J Surg*. 1989; **76**: 339–41.
30 Silverman R E, Mcleod R S, Cohen Z. Strictureplasty in Crohn's disease. *Can J Surg*. 1989; **32**: 19–22.
31 Sayfan J, Wilson D A L, Allan A, *et al*.. Recurrence after strictureplasty or resection for Crohn's disease. *Br J Surg*. 1989; **76**: 335–8.
32 Kendall G P N, Hawley P R, Nicholls R J, Lennard-Jones J E. Strictureplasty:

a good operation for small bowel Crohn's disease? *Dis Colon Rectum*. 1986; **29**: 312–6.

33 Davis J M. The prognosis of Crohn's disease of the small intestine. *Postgrad Med J*. 1961; **37**: 783–91.

34 Hamilton S R, Boitnott J K, Morson B C. Relationships of disease extent and margin lengths to recurrence of Crohn's disease after ileocolonic anastomosis. *Gastroenterology*. 1981; **80**: 1166.

35 Trnka Y M, Glotzer D J, Kasdon E J, *et al.*. Long-term outcome of restorative operation in Crohn's disease. *Ann Surg*. 1982; **196**: 345–55.

36 Opelz G, Sengar D P S, Mickey M R, Terasaki P I. Effect of blood transfusions on subsequent kidney transplants. *Transplant Proc*. 1973; **5**: 253–9.

37 Hutchinson I V, Morris P J. The role of major and minor histocompatibility antigens in active enhancement of rat kidney allograft survival by blood transfusion. *Transplantation*. 1986; **41**: 166–70.

38 Veitch P S, Shenton B K, Proud G, Taylor R M. Plasma suppressive activity and kidney graft survival. *Br J Surg*. 1980; **67**: 703–707.

39 Keown P A, Descamps B. Improved renal allograft survival after blood transfusion: a non-specific, erythrocyte-mediated immunoregulatory process? *Lancet*. 1979; **1**: 20–22.

40 Terasaki P I. The beneficial transfusion effect on kidney graft survival attributed to clonal deletion. *Transplantation*. 1984; **37**: 119–25.

41 Lenhard V, Maassen G, Grosse-Wilde H, *et al.*. Effect of blood transfusions on immunoregulatory mononuclear cells in prospective transplant recipients. *Transplant Proc*. 1983; **15**: 1011–5.

42 Takeuchi H, Sakagami K, Seki Y, *et al.*. Anti-idiotypic antibodies and suppressor cells induced by donor-specific transfusion in potential kidney transplant recipients. *Transplant Proc*. 1985; **17**: 1059–61.

43 Singal D P, Fagnilli L, Joseph S. Blood transfusions induce antiidiotypic antibodies in renal transplant patients. *Transplant Proc*. 1983; **15**: 1005–1008.

44 Macleod A, Catto G, Mather A, *et al.*. Beneficial antibodies in renal transplantation developing after blood transfusion: evidence for HLA linkage. *Transplant Proc*. 1985; **17**: 1057–8.

45 Burrows L, Tartter P. Effects of blood transfusions on colonic malignancy recurrence rates. *Lancet*. 1982; **2**: 662 (Letter).

46 Blumberg N, Agarwal M M, Chuang C. Relation between recurrence of cancer of the colon and blood transfusion. *Br Med J*. 1985; **290**: 1037–9.

47 Blumberg N, Heal J M. Perioperative blood transfusion and solid tumor recurrence – a review. *Cancer Invest*. 1987; **5**: 615–25.

48 Foster R S, Jr, Costanza M C, Foster J C, *et al.*. Adverse relationship between blood transfusions and survival after colectomy for colon cancer. *Cancer*. 1985; **55**: 1195–201.

49 Tartter P I, Quintero S, Barron D M. Perioperative blood transfusion associated with infectious complications after colorectal cancer operations. *Am J Surg*. 1986; **152**: 479–82.

50 Voogt P J, Van de Velde C J H, Brand A, *et al.*. Perioperative blood transfusion and cancer prognosis. Different effects of blood transfusion on prognosis of colon and breast cancer patients. *Cancer*. 1987; **59**: 836–43.

51 Heal J M, Chuang C, Blumberg N. Perioperative blood transfusions and prostate cancer recurrence and survival. *Am J Surg*. 1988; **156**: 374–80.

52 Rosenberg S A, Seipp C A, White D E, Wesley R. Perioperative blood transfusions are associated with increased rates of recurrence and decreased survival in patients with high-grade soft-tissue sarcomas of the extremities. *J Clin Oncol*. 1985; **3**: 698–709.

53 Hyman N H, Foster R S, Jr, DeMeules J E, Costarya M C. Blood transfusion and survival after lung cancer resection. *Am J Surg*. 1985; **149**: 502–7.
54 Tartter P I, Burrows L, Papatestas A E. Peri-operative blood transfusion has prognostic significance for breast cancer. *Surgery*. 1985; **97**: 225–30.
55 Blumberg N, Agarwal M, Chuang C. A possible association between survival time and transfusion in patients with cervical cancer. *Blood*. 1985; **66** (II): 274a (Abstr).
56 Gatti R A, Good R A. Occurrence of malignancy in immunodeficiency diseases: a literature review. *Cancer*. 1971; **28**: 89–98.
57 Kersey J H, Spector B D, Good R A. Cancer in children with primary immunodeficiency diseases. *J Pediatr*. 1974; **84**: 263–4.
58 Kinlen L J, Sheil A G R, Peto J, Doll R. Collaborative United Kingdom-Australasian study of cancer in patients treated with immunosuppressive drugs. *Br Med J*. 1979; **2**: 1461–6.
59 Birkeland S A. Malignant tumors in renal transplant patients. The Scandia transplant material. *Cancer*. 1983; **51**: 1571–5.
60 Peters W R, Fry R D, Fleshman J W, *et al.*. Multiple blood transfusions reduce the recurrence rate of ileocolic Crohn's disease. Presented at the meeting of *The American Society of Colon and Rectal Surgeons*: Anaheim, California, 1988.
61 Peters W R, Fry R D, Fleshman J W, Kodner I J. Multiple blood transfusions reduce the recurrence rate of Crohn's disease. *Dis Colon Rectum*. 1989; **32**: 749–53.
62 Williams J G, Hughes L E. Effect of perioperative blood transfusion on recurrence of Crohn's disease. *Lancet*. 1989; **2**: 131–3.
63 Sachar D B, Taub R N, Brown S M, *et al.*. Impaired lymphocyte responsiveness in inflammatory bowel disease. *Gastroenterology*. 1973; **64**: 203–209.
64 Auer I O, Wechsler W, Ziemer E, *et al.*. Immune status in Crohn's disease. I: Leukocyte and lymphocyte subpopulations in peripheral blood. *Scand J Gastroenterol*. 1978; **13**: 561–71.
65 Verrier-Jones J, Housley J, Ashurst P M, Hawkins C F. Development of delayed hypersensitivity to dinitrochlorobenzene in patients with Crohn's disease. *Gut*. 1969; **10**: 52–6.
66 Francis D M A, Judson R T. Blood transfusion and recurrence of cancer of the colon and rectum. *Br J Surg*. 1987; **74**: 26–30.
67 Cheslyn-Curtis S, Fielding L P. Blood transfusion and cancer recurrence. *Surgery*. 1988; **62**: 1468–70.
68 Francis D M A, Shenton B K, Proud G, Taylor R M R. Immunosuppressive plasma factors in malignant disease. *Aust N Z J Surg*. 1985; **55**: 111–20.
69 Chambers T J, Morson B C. The granuloma in Crohn's disease. *Gut*. 1979; **20**: 269–74.
70 Glass R E, Baker W N W. Role of the granuloma in recurrent Crohn's disease. *Gut*. 1976; **17**: 75–7.
71 Wolfson D M, Sachar D B, Cohen A, *et al.*. Granulomas do not affect postoperative recurrence rates in Crohn's disease. *Gastroenterology*. 1982; **83**: 405–409.
72 Heimann T M, Miller F, Martinelli G, *et al.*. Correlation of presence of granulomas with clinical and immunologic variables in Crohn's disease. *Arch Surg*. 1988; **123**: 46–8.
73 Hermanowicz A, Gibson P R, Pallone F. The function *in vitro* of macrophages from the intestinal mucosa of patients with Crohn's disease: an association between chemotactic migration and granulomata. *J Gastroenterol Hepatol*. 1988; **3**: 117–25.
74 Ward M. The pathogenesis of Crohn's disease. *Lancet*. 1977; **2**: 903–905.

75 Scammell B, Ambrose N S, Alexander-Williams J, *et al.*. Recurrent small bowel Crohn's disease is more frequent after subtotal colectomy and ileorectal anastomosis than proctocolectomy. *Dis Colon Rectum*. 1985; **28**: 770–71.
76 Burman J H, Thompson H, Cooke W T, Alexander-Williams J. The effects of diversion of intestinal contents on the progress of Crohn's disease of the large bowel. *Gut*. 1971; **12**: 11–15.
77 Harper P H, Truelove S C, Lee E C, *et al.*. Split ileostomy and ileocolostomy for Crohn's disease of the colon and ulcerative colitis: a 20-year survey. *Gut*. 1983; **24**: 106–113.
78 Longo W E, Ballantyne G H, Cahow E. Treatment of Crohn's colitis. Segmental or total colectomy? *Arch Surg*. 1988; **123**: 588–90.
79 Hamilton S R, Reese J, Pennington L, *et al.*. No role of resection margin frozen sections in the surgical management of Crohn's disease. *Gastroenterology*. 1982; **82**: 1078 (Abstr).
80 Alexander-Williams J, Fielding J F, Cooke W T. A comparison of results of excision and bypass for ileal Crohn's disease. *Gut*. 1972; **13**: 973–5.
81 Goligher J C. Inflammatory disease of the bowel. Results of resection for Crohn's disease. *Dis Colon Rectum*. 1976; **19**: 584–7.
82 Farmer R G, Hawk W A, Turnbull R B. Clinical patterns in Crohn's disease. A statistical study of 615 cases. *Gastroenterology*. 1975; **68**: 627–35.
83 Flint G, Strauss R, Platt N, Wise L. Ileorectal anastomosis in patients with Crohn's disease of the colon. *Gut*. 1977; **18**: 236–9.
84 Koch T R, Cave D R, Ford H, Kirsner J B. Crohn's ileitis and ileocolitis. A study of the anatomical distribution of recurrence. *Dig Dis Sci*. 1981; **26**: 528–31.
85 Wakefield A J, Dhillon A P, Rowles P M, *et al.*, Pathogenesis of Crohn's disease: multifocal gastrointestinal infarction. *Lancet*. 1989; **2**: 1057–62.
86 Berman S, DuPree J, Matsumoto T, *et al.*. Intraoperative hemostasis and wound healing in intestinal anastomoses using the ILA stapling device. *Am J Surg*. 1988; **155**: 520–25.
87 Ferrara J J, Dyess D L, Lasecki M, *et al.*. Surface oximetry. A new method to evaluate intestinal perfusion. *Am Surg*. 1988; **54**: 10–14.
88 Roberts J O, Jones B M. Direct anastomotic assessment for postoperative microvascular monitoring. *Ann Plastic Surg*. 1987; **19**: 219–24.
89 Gleeson M J, McMullin J P. Suture granuloma simulating a cholangiocarcinoma. *Br J Surg*. 1987; **74**: 1181.
90 Phillips R K S, Cook H T. Effect of steel wire sutures on the incidence of chemically-induced rodent colonic tumours. *Br J Surg*. 1986; **73**: 671–4.
91 McGregor J R, Galloway D J, Jarrett F, George W D. Anastomotic materials and colorectal carcinogenesis. *Br J Surg*. 1988; **75**: 603 (Abstr).

11

Faecal incontinence: an American perspective

J. Graham Williams and David A. Rothenberger

Introduction

There are few benign conditions that are as devastating to the afflicted individual as faecal incontinence. Patients with severe incontinence are often house bound and even patients with minor incontinence experience disruption in their lifestyle. The prevalence of incontinence in the community is unknown and depends, in part, on how incontinence is defined. One community-based survey estimated the prevalence to be 4.2/1000 males between 15 and 65 years of age, rising to 10.9/1000 for those over 65 years of age.[1] Faecal incontinence is common in the elderly with a prevalence rate of 17 per cent reported in a group of nursing home residents.[2] However, the true prevalence is probably higher. In our practice we see many patients with minor anorectal conditions who admit to episodes of anal incontinence only after direct questioning.

Anal continence

Despite extensive study, the exact mechanism of anal continence is incompletely understood. Continence depends on a complex interplay between a number of factors (Table 11.1). The volume and consistency of stool arriving in the rectum is important. Rapid colonic transit or increased fluid content of stool, resulting from either decreased absorption or increased secretion, may result in incontinence, even in the presence of a normal pelvic floor and anal sphincter. Similarly, a compliant rectal reservoir is necessary for normal continence. Patients with reduced rectal compliance, as may be seen in inflammatory bowel disease, ischaemic colitis or following pelvic irradiation, suffer with rectal urgency.[3,4]

The ability to sense the presence of stool in the rectum and to discriminate between gas, liquid or solid stool is fundamental to maintaining faecal continence. The receptor mechanism involved in recognizing rectal disten-

Table 11.1 Factors important in maintaining faecal continence

Intestinal transit
Stool volume and consistency
Rectal capacity and compliance
Rectal motility
Anorectal sensation
Anorectal angle
Anal sphincter tone
Neuromuscular coordination

sion is thought to lie in the muscles of the pelvic floor, rather than the rectal wall. This explains why patients who have undergone rectal excision and colo-anal anastomosis or an ileoanal pouch procedure retain the ability to sense the presence of stool.[5,6] The other important sensory area lies in the transitional zone of the anal canal.[7] During rectal distension, there is simultaneous relaxation of the internal sphincter and contraction of the external sphincter (the recto-anal inhibitory reflex).[8] This allows rectal contents to descend into the sensory-rich upper anal canal, whilst external sphincter contraction occludes the distal anal canal and prevents incontinence.[9]

Impaired anorectal sensation is an important cause of incontinence. Elderly patients with incontinence caused by faecal impaction have been shown to have diminished rectal sensation and are unable to perceive the presence of a bolus of stool in the rectum and seepage of liquid stool into the anal canal.[10,11] Similarly, diabetic patients who are incontinent of faeces have been shown to have impaired rectal sensation.[12] Diminished anal sensation appears to be an important contributing factor in incontinence in patients with descending perineum syndrome[13] and in patients with neurogenic incontinence.[14]

The high-pressure zone of the anal canal results from activity of the internal sphincter, external sphincter and the puborectalis muscles. 85 per cent of resting tone is thought to be produced by tonic contraction of the internal sphincter and the remainder by the external sphincter.[19–21] Activity in the external sphincter increases with erect posture, coughing and increased intra-abdominal pressure.[16] Voluntary contraction pressure is greatest in the distal anal canal where the mass of external sphincter muscle is greatest.

The importance of the anorectal angle in the continence mechanism is controversial. The anorectal angle is formed by the anteriorly directed pull of the puborectalis sling. In normal subjects, this angle varies from 60°–105° at rest and becomes more acute during contraction of the sphincters.[15] During defaecation, the angle straightens, allowing the rectum to empty. Parks proposed that the angle was important, acting as a flap valve to seal the anus when intra-abdominal pressure increased.[16] A number of studies have shown that the anorectal angle is more obtuse in incontinent patients, suggesting that the efficiency of the 'flap valve' is reduced.[16,17] However,

Bartolo *et al.*[18] showed that the rectal wall is always separated from the anal canal in normal subjects during a maximal Valsalva manoeuvre and thus cast doubt on the flap valve theory.

The relative importance of the anorectal angle versus sphincter tone in maintaining continence is controversial. The internal sphincter can often be divided in its entirety without compromising continence, and it has been known for some time that division of the external sphincter does not necessarily result in incontinence, provided the puborectalis sling is preserved.[22] However, a number of studies have cast doubt on the flap valve theory of continence and have shown that sphincter-produced anal canal pressure always exceeds rectal pressure during valsalva manoeuvre in continent subjects.[18,23]

Aetiology of faecal incontinence

A classification of the causes of faecal incontinence is shown in Table 11.2. In our practice, the most common causes of faecal incontinence are obstetric injuries, neurogenic incontinence and incontinence in association with rectal prolapse.

Table 11.2 Causes of faecal incontinence

Congenital	**Secondary**
Spina bifida	Diarrhoea
Imperforate anus	Inflammatory bowel disease
Hirschprung's disease	Short gut
Others	Infective diarrhoea
	Laxative abuse
Traumatic	
Impalement, pelvic fracture	**Diminished rectal reservoir**
Surgical	Irradiation
Fistulotomy	Inflammatory bowel disease
Haemorrhoidectomy	Anterior resection
Sphincterotomy/stretch	
Obstetric	**Overflow**
Tear	
Episiotomy	**Systemic and neurological disease**
	Neuropathy (diabetic)
Neurogenic	Scleroderma
Aging	Multiple sclerosis
Pudendal neuropathy	Dementia

Obstetric incontinence

The close proximity of the anus to the vagina places the anal sphincters at risk of damage during childbirth by extension of a perineal laceration or a midline episiotomy to include the sphincters, from an overzealous episiotomy, particularly when performed in the midline, or from a large perineal haematoma or infection. Most American obstetricians prefer to use a midline episiotomy, which may increase the risk of sphincter injury especially in the 10 per cent of women who have an anteriorly displaced anus. Most injuries are recognized at the time of delivery and are repaired primarily. Laxity of the pelvic floor and good blood supply usually permit a satisfactory repair. However, the repair may break down because of complications such as sepsis, haematoma and suture breakage, resulting in sphincter disruption and/or a recto-vaginal fistula. In some patients, damage to the sphincter is not recognized at the time of delivery and only becomes apparent when the patient develops faecal incontinence at a later date.

Neurogenic incontinence

Neurogenic incontinence mainly affects post-menopausal women. There is an association with a history of difficult childbirth or in nulliparous women with constipation and straining at stool. On examination, the anus is usually patulous and when the patient strains, the perineum descends.[24] Anorectal manometry shows decreased resting pressures and markedly reduced squeeze pressures.[25] Anal canal sensation is also reduced.[26,27] The pathophysiology of neurogenic incontinence is thought to involve progressive denervation of the pelvic floor and anal sphincters. Histological studies of biopsies of puborectalis, levator ani and external anal sphincter have shown evidence of chronic partial denervation.[28] Electrophysiology studies, using concentric needle or single fibre electromyographic techniques, have shown severe abnormalities in patients with neurogenic incontinence, with evidence of compensatory re-innervation.[29,30] The neural supply of the external sphincter has also been shown to be impaired by measuring terminal motor latency in the pudendal nerve.[31]

The neuropathic changes in the pelvic floor and anal sphincter are thought to be caused by stretch injury to the pudendal nerve and the nerves supplying the puborectalis during childbirth or by prolonged straining. Pudendal nerve terminal motor latency is more prolonged in nulliparous women than in primiparous women.[32] Prolonged labour, forceps delivery and a large baby are all factors that are associated with an increased degree of nerve damage.[32,33] Straining has been shown to increase pudendal nerve terminal motor latency in patients with perineal descent.[34] Furthermore, a linear relationship between the degree of perineal descent and pudendal nerve terminal motor latency has been demonstrated.[35]

Rectal prolapse

Faecal incontinence occurs in about 50 per cent of patients with rectal prolapse.[36] Incontinent patients with rectal prolapse have lower resting anal pressures and lower maximum voluntary contraction pressures than continent patients with rectal prolapse.[36–40] Factors thought to be important in the aetiology of incontinence in association with rectal prolapse include: stretching of the anal sphincters,[41] chronic inhibition of the internal sphincter via the rectoanal inhibitory reflex,[42] and loss of anorectal sensation.[41] Damage to the innervation of the pelvic floor and anal sphincters is probably the most important factor in the aetiology of incontinence in rectal prolapse. Work from St Mark's Hospital has demonstrated histological and electromyographic evidence of denervation of the pelvic floor and external anal sphincter in patients with rectal prolapse who were incontinent.[25,28,43]

A number of studies have attempted to explain why continence improves following prolapse repair. Increased resting anal pressure following prolapse repair has been observed by a number of authors,[36,42] but has not been a universal finding.[38–40] Similarly, increased maximum voluntary contraction pressure has been demonstrated,[36] but not by most investigators.[38,39,44] Bartolo and Duthie[45] were unable to demonstrate any change in resting anal pressure or maximum voluntary contraction pressure following prolapse repair, but did show a significant improvement in rectal sensation and sensation in the upper anal canal.

A number of investigators have attempted to identify those incontinent patients with rectal prolapse who would not improve following prolapse repair, by using preoperative anorectal physiology studies. In our laboratory, we studied 14 incontinent patients with rectal prolapse, of whom five did not improve or regain continence following prolapse repair. These five patients had significantly lower preoperative resting anal pressure and maximum voluntary contraction pressure than the other nine patients who did improve.[36] However, the Birmingham group did not show any predictive value for preoperative resting anal pressure or maximum voluntary contraction pressure. They suggested that a short physiological anal canal, low volume at first leak during a saline infusion test and greater pelvic floor descent during straining were predictors of persisting incontinence.[46]

In some patients, incontinence results from a number of different causes acting together, such as in Crohn's disease, where a weak sphincter is unable to control loose stool. In a proportion of patients with an obstetric tear of the sphincter, there is also evidence of pudendal neuropathy.[47]

Assessment of the patient

Complete evaluation of the patient with incontinence is required for an accurate diagnosis to be made, and for the appropriate treatment to be offered.

History

A full history is required, which should establish whether the patient has true incontinence or a degree of incontinence and the likely aetiology of incontinence. A patient may think they are incontinent, but in reality have peri-anal soiling as a result of mucosal or haemorrhoidal prolapse, poor anal hygiene, a perianal fistula, anorectal venereal disease or an anorectal neoplasm. A full medical history is required and in females a careful obstetric history should be taken, including the use of forceps, perineal trauma and episiotomy and any repairs performed. The degree of incontinence can be quantitated by using an incontinence score. We favour the method described by Miller *et al.*,[48] which gives a numerical score for incontinence to gas, liquid and solid stool, based on frequency of incontinence and we also take into account the degree to which incontinence affects the patient's life style. This scoring system is useful for assessing objectively the effect of treatment on incontinence.

Examination

Careful physical examination should be performed. During inspection, particular attention should be paid to the presence of peri-anal soiling, a patulous anus, scars, mucosal or haemorrhoidal prolapse and the presence of a fistula. The patient should be asked to strain down to exclude a rectal prolapse and to assess the degree of perineal descent. It is best to have the patient squat or sit on a commode while straining to defaecate to exclude procidentia. Careful digital examination should assess both resting tone and squeeze pressure, as well as puborectalis movement. Anal canal compliance should be assessed as well as the rectal contents. Palpation of the anus may reveal a defect in the sphincter or soft tissue scarring.

Investigations

Following a full history and examination it is usually possible to make an accurate diagnosis as to the cause of incontinence. However, a number of investigations are necessary to confirm the diagnosis and to obtain baseline measurements of various parameters.

Endoscopy should be performed, by either rigid or flexible sigmoidoscopy, to exclude a neoplasm or mucosal inflammation and to look for evidence of a solitary ulcer or rectal intussusception. We study the majority of our incontinent patients in the anorectal physiology laboratory. Anorectal manometry is performed using a four lumen concentric catheter by an open, water-perfused technique. The data are recorded on a computer that calculates the mean resting pressure and maximum voluntary contraction and squeeze pressure at each level in the anal canal. Rectal sensation is assessed by distending a balloon in the rectum and recording the volume of air infused when the patient first perceives the presence of the balloon and the maximum tolerated volume. In patients in whom neurogenic incontinence is suspected, pudendal nerve latency is measured using the method described by St Mark's Hospital. We also measure pudendal nerve latency in patients with obstetric tears to exclude an associated neuropathy. Mapping of the external sphincter using needle electromyography is performed occasionally in patients with traumatic sphincter injuries. Video defaecography is performed in certain patients to measure the anorectal angle, assess the degree of perineal descent and to look for an internal prolapse.

Treatment of faecal incontinence

Non-surgical treatment

Many patients with minor incontinence can be managed successfully by non-surgical measures. The simplest method is to establish a bowel management programme with a routine for complete evacuation of the rectum, to ensure that overflow incontinence does not occur. Dietary manipulation is required, which usually involves limiting milk products and increasing the amount of fibre and ensuring an adequate water intake. Stimulant laxatives, suppositories and daily enemas are sometimes required to achieve adequate rectal emptying. Diarrhoea, if present, should be investigated and the underlying disease or infectious agent treated. Antimotility drugs are useful in patients with diarrhoea resulting from rapid intestinal transit. Loperamide has been shown to improve continence, increase anal sphincter pressures, decrease stool volume, and increase stool consistency.[49] Occasionally, an anatomical problem such as a rectocele can be managed by encouraging the patient to support the anterior rectal wall during defaecation with a finger in the vagina, which will aid complete evacuation of the rectum and prevent later leakage of stool and mucus.

Biofeedback

Biofeedback training is used in our institution to treat selected patients with faecal incontinence. Biofeedback requires an instrument to provide information to the patient about a physiological process that is not accurately perceived by the patient, so that they may learn how to control the process. Biofeedback only affects existing responses. For treating faecal incontinence, the patient is provided with a stimulus (rectal distension) and a recording apparatus that reflects anal sphincter function. This provides the patient with a measure to determine whether sphincter contraction is performed appropriately following rectal distension.[50] Biofeedback is appropriate treatment for a number of causes of faecal incontinence, both as primary treatment or following surgery. Patients suitable for biofeedback should be well motivated, be able to understand and process instructions, retain the ability to contract their anal sphincter and have adequate rectal sensation. Various systems are available to perform biofeedback. These include triple or double balloon systems that measure sphincter responses manometrically and intra-anal electrodes that measure the strength of electrical activity in the anal sphincter. Sphincter responses are fed back to the patient by a visual display or auditory signal.

The reported results of biofeedback for faecal incontinence are summarized in Table 11.3. Success of biofeedback depends in part on what definition is used. Most series include a variety of causes of incontinence. The

Table 11.3 Results of biofeedback for faecal incontinence

Author	Date	n	Follow-up	Success definition	Success number (%)
Cerulli *et al.*[51]	1979	50	4–108 months	90% ↓ in soiling	30 (72%)
Goldenberg *et al.*[52]	1980	12	3–24 months	Not stated	10 (83%)
Wald[53]	1981	17	2–38 months	75% ↓ in soiling	12 (70%)
Wald & Tunuguntla[12]	1984	11	7–24 months	Not stated	8 (73%)
Whitehead *et al.*[54]	1985	13	< 12 months	75% ↓ in soiling	10 (77%)
MacCleod[55]	1987	113	6–60 months	90% ↓ in soiling	71 (63%)
Riboli *et al.*[56]	1988	21	Not stated	90% ↓ in soiling	18 (86%)
Jensen and Lowry[57]	1991	31	> 6 months	90% ↓ in soiling	28 (90%)

likelihood of success appears to be related to the aetiology of incontinence. MacLeod[55] noted that biofeedback was most effective in postobstetric and certain postsurgical cases, but was not helpful where a keyhole deformity of the anus was present or following anterior resection of the rectum. The most important factor in the success of biofeedback remains uncertain. It is likely that the feedback the patient receives is important, as sphincter exercises alone are of limited value.[54] Manometric studies have shown only minimal improvement in squeeze pressures after biofeedback.[56,58] However, improvements in the threshold of rectal sensation appear to be important in the success of treatment.[51,58,59]

Thus biofeedback is a useful treatment for faecal incontinence in selected patients. Further studies should define which patients are best treated by biofeedback and the most efficient method of performing the technique.

Surgical treatment

A number of surgical approaches have been developed for treating faecal incontinence. These can be broadly divided into operations to repair or augment the anal sphincters and pelvic floor, operations which correct an underlying disease or condition and palliative operations (Table 11.4).

Table 11.4 Surgical approaches to faecal incontinence

Sphincter repair/augmentation	Correction of the underlying disease
Primary repair	*Rectal prolapse*
Secondary repair	Transabdominal repair
Sphincteroplasty	Perineal repair ± levatoroplasty
Overlapping ± levatoroplasty	*Carcinoma*
Plication	Restorative resection
Post-anal repair	*Radiation*
Encirclement	Restorative resection
Foreign material	*Inflammatory bowel disease*
Theirsch procedure	Restorative resection
Modified Theirsch	
Artificial anal sphincter	
Muscle transfer	**Palliative**
Gracilis or other muscles	*Stoma*
Free-muscle transfer	End
Stimulated-muscle transfer	Loop

Sphincter repair

Primary repair

Acute, traumatic sphincter injuries caused by impalement, obstetric laceration etc., are best treated by primary repair whenever possible. Direct apposition of the severed ends of muscle, using 2/0 polyglycolic acid mattress sutures produces satisfactory results. Extensive mobilization of the muscle is not required. A defunctioning colostomy is not required for a simple laceration.

If extensive contamination is present, debridement of devitalized tissue is necessary. Primary sphincter repair can still be performed, but the skin should be left open and a defunctioning colostomy performed. If other life-threatening injuries are present, or there is extensive local trauma, a defunctioning colostomy is performed and sphincter repair is delayed.

Secondary repair

Secondary repair is usually performed when a primary repair fails because of sepsis, haematoma or faulty technique. In a number of patients, the sphincter injury may not have been noticed at the time of trauma. Secondary repair is performed when all wounds have healed and inflammation has resolved. This generally takes at least 3 months.

The technique we use has been described in detail elsewhere[60]. A defunctioning colostomy is not performed. The patient is prepared with full mechanical bowel preparation and prophylactic antibiotics. General anaesthetic is administered and the bladder drained with a Foley catheter. The operation is performed in the prone jackknife position, with the aid of headlights for adequate illumination. A curved incision is made in the region of the sphincter defect. The anoderm and rectal mucosa are mobilized off the scar tissue and the divided ends of muscle are identified. The sphincter is mobilized from scar and surrounding tissues. For major injuries, we extend the dissection to expose the levator muscles. If a rectovaginal fistula is present, this can be repaired at the same time as sphincter repair.

The technique of sphincter reconstruction depends on the severity of the injury. A distal sphincter injury, with complete sphincter division and a thin bed of scar tissue is treated by overlapping sphincteroplasty, with four to eight horizontal mattress sutures of 2/0 polyglycolic acid. Where the entire sphincter has been divided, the anterior levator muscles are plicated prior to performing an overlapping sphincteroplasty. The aim of the levatoroplasty is to strengthen and lengthen the anal canal. For less severe injuries, where the sphicter is not completely divided but is attenuated, we plicate the intact muscle and scar tissue with horizontal mattress sutures.

Following sphincter repair, the wounds are left partly open and gently packed. Strong analgesia is required and an epidural infusion of an opiate is a useful technique. The patient is kept on clear liquids until the first bowel movement, when solid food is resumed. Bulk laxatives and daily enemas are administered to ensure that faecal impaction does not develop.

Published results of secondary sphincter repair are summarized in Table 11.5. A good result can be expected in about 80 per cent of patients. A number of factors have been shown to influence the chance of a successful outcome. Sphincter repair after obstetric injury has been shown to be associated with better results than repair of iatrogenic sphincter division.[63,69] A poorer outcome has been noted in older patients than in younger patients.[69] In older patients, however, other factors may play a role in the

Table 11.5 Results of secondary sphincter repair for incontinence

Author	Date	Centre	n	Aetiology of injury*			Results		
				Obstetric (%)	Trauma (%)	Operative (%)	Good (%)	Fair (%)	Poor (%)
Castro & Pittman[61]	1978	Maryland	18	N/S	N/S	N/S	15 (83)	0	3 (17)
Keighley & Fielding[62]	1983	Birmingham	19	2 (11)	4 (21)	11 (57)	11 (57)	5 (26)	3 (16)
Browning & Motson[63]	1984	St Mark's	97	13 (13)	26 (26)	58 (58)	75 (78)	13 (13)	9 (9)
Fang *et al.*[64]	1984	Univ. of Minnesota	79	43 (54)	7 (9)	27 (34)	70 (89)	6 (7)	3 (4)
Pezim *et al.*[65]	1987	Mayo Clinic	40	23 (58)	0	17 (42)	13 (33)	20 (50)	7 (17)
Yoshioka & Keighley[66]	1989	Birmingham	27	9 (33)	4 (15)	11 (41)	7 (26)	13 (48)	7 (26)
Miller *et al.*[67]	1989	Bristol	30	14 (47)	16 Neurogenic		20 (67)	6 (20)	4 (13)
Wexner *et al.*[68]	1991	Florida	16	16 (100)	–	–	12 (75)	3 (19)	1 (6)

*Most series included other miscellaneous types of sphincter injury. N/S = not stated. Definition of result: Good, continent of solid and liquid stool; Fair, continent of solid stool only with leakage of liquid; Poor, incontinent of solid and liquid stool.

aetiology of incontinence, such as unrecognized neuropathy. A number of authors have noted that the results are poorer in patients who have undergone previous attempts at sphincter repair.[63,69] Concomitant repair of a recto-vaginal fistula does not appear to affect the outcome of sphincter repair.[64,69] The influence of postoperative sepsis on the outcome of sphincter repair is unclear. Keighley and Fielding[62] noted that all three patients in their series who developed severe infection after sphincter reconstruction had persisting incontinence and other authors have noted a poorer outcome if sepsis develops.[61,63,69] For this reason, some surgeons have stressed the importance of a covering colostomy.[62] However, most American surgeons, including ourselves, consider a routine covering colostomy to be unnecessary, whilst British and Australian surgeons are more likely to perform a covering colostomy.[62] This may be related to differences in the aetiology of sphincter injury, with obstetric tears of the sphincter more common in American practice.

A poor result can be expected from repair of an obstetric tear when there is damage to the nerve supply of the sphincter.[63,71] Snooks *et al.*[32] showed physiological evidence of damage to the innervation of the external sphincter in 60 per cent of patients with an obstetric injury and faecal incontinence. Jacobs *et al.*[71] demonstrated severe denervation by electromyography in all five of their patients who remained incontinent after sphincter repair, whereas 12 patients with a successful repair showed no evidence of neural damage. Similar results were reported by Laurberg *et al.*,[47] who obtained excellent results in eight of ten patients with obstetric tears who had no evidence of nerve damage and only one of nine patients who had evidence of nerve damage. However, Miller *et al.*[67] reported restoration of continence to near normal in ten of 16 patients with neurogenic incontinence treated by anterior sphincter repair and levatoroplasty.

The effect of sphincter repair on anorectal physiology has been investigated by a number of groups. Keighley and Fielding[62] reported a significant increase in maximum voluntary contraction pressure after successful sphincter repair, but no change in resting anal pressure. Maximum voluntary contraction pressure did not increase in three patients who remained incontinent. Similar results were reported by Miller *et al.*,[67] who combined anterior levatoroplasty with sphincter repair. Furthermore, they demonstrated an improvement in anal canal sensation, which they postulated was important in improvement in continence. Roberts *et al.*[72] were only able to demonstrate an increase in resting anal pressure after sphincter repair. However, only four patients were studied manometrically pre- and postoperatively. Wexner *et al.*[68] demonstrated significant increase in resting anal pressure, maximum voluntary contraction pressure and length of the high-pressure zone of the anal canal in 16 patients undergoing anterior sphincteroplasty for obstetric injury.

Post-anal repair

The operation of post-anal repair was devised by Sir Alan Parks.[16] The operation is based on the theory that an acute angle between the rectum and anal canal is important in maintaining continence by a 'flap valve' mechanism.[16] Many patients with faecal incontinence have decreased anal pressures, decreased sphincter length and an obtuse anorectal angle. Post-anal repair was designed to restore the anorectal angle and increase sphincter length by plication of the pelvic floor muscles. Post-anal repair is indicated in patients with neurogenic incontinence, patients with obstetrical injuries if the sphincter remains intact but the pelvic floor innervation has been damaged by prolonged labour, patients who remain incontinent after imbrication sphincteroplasty and patients with rectal prolapse who remain incontinent after rectopexy.

The operative technique has been described in detail elsewhere.[73] Briefly, the patient is prepared as for sphincteroplasty. We favour the prone jackknife position, but the operation can be performed in lithotomy position if desired. A 'V' shaped or transverse incision is made posterior to the anus, with the apex towards the coccyx. The intersphincteric plane is identified and dissection continues cephalad in this relatively avascular plane, in the posterior midline, until the puborectalis levator sling is reached. The supralevator space is entered by dividing the rectosacral fascia. The limbs of the levators are identified and are approximated by a series of 2/0 polypropylene horizontal mattress sutures. The sutures are tied without tension to avoid ischaemia of the muscles. More distally the pubococcygeus and puborectalis muscles are approximated and plicating sutures are placed in the external sphincter to narrow the anal canal. The incision is closed over a small suction drain.

The reported results of post-anal repair are summarized in Table 11.6. Some authors, most notably Parks,[16] have reported excellent results after this operation, with restoration of continence in about 80 per cent of patients. However, other groups have not achieved such a high success rate and we have been disappointed with our results from this operation. The length of follow-up may be important when assessing the results of

Table 11.6 Results of post-anal repair for faecal incontinence

Author	Date	n	Follow-up	Cont (%)	Impr (%)	Fail (%)
Parks[16]	1975	75	< 15 years	62 (83)	N/S –	13 (17)
Keighley & Fielding[62]	1983	40	N/S	27 (67)	34 (85)	6 (15)
Browning & Parks[74]	1983	42	N/S	34 (81)	N/S –	8 (19)
Henry & Simson[75]	1985	204	1–27 months	119 (58)	N/S –	61 (30)
Yoshioka *et al.*[76]	1987	79	> 3 years	24 (30)	50 (69)	29 (31)
Womack *et al.*[77]	1988	16	15–48 months	6 (38)	14 (88)	2 (13)
Miller *et al.*[48]	1988	17	3–40 months	10 (59)	13 (76)	4 (24)
Yoshioka *et al.*[78]	1988	19	6 months	N/S –	12 (63)	7 (37)
Scheur *et al.*[79]	1989	39	N/S	17 (43)	27 (69)	12 (31)
Rainey *et al.*[80]	1990	42	6–95 months	13 (31)	29 (69)	12 (29)

Cont, fully continent after surgery with only occasional loss of flatus; Impr, continence improved after surgery, includes patients who became fully continent; Fail, patients who had no change in degree of incontinence. N/S, not stated.

this operation. Keighley and Fielding[62] described restoration of full continence in 27 of 40 patients (67 per cent) treated by post-anal repair. However, review of 79 patients followed up for longer than 3 years by the same group revealed that only 24 patients (30 per cent) were completely continent.[76]

The exact mechanism for restoration of continence by this operation remains controversial. The success of the operation was initially thought to result from restoration of an acute anorectal angle.[16] However, a number of physiological studies have failed to demonstrate any change in the anorectal angle following successful post-anal repair,[48,77,78] casting doubt on the importance of the anorectal angle and flap valve in maintaining continence. A number of studies have demonstrated significant increases in resting anal pressure and maximum voluntary contraction pressure following successful post-anal repair.[48,74] However, other groups have found no change in sphincter pressures.[77,78] Anal canal sensation improves following post-anal repair[48] and the length of the physiological anal canal increases,[74,77] which may be important in restoration of continence. Physiological parameters that have been shown to be predictive of a poor outcome from post-anal repair include low resting and maximum voluntary contraction pressures and marked perineal descent.[78] Others have reported a good result in patients with low preoperative sphincter pressures.[79]

Thus post-anal repair does appear to benefit a proportion of patients with neurogenic incontinence, although most authors have been unable to reproduce Sir Alan Parks' results. It appears that the operation lengthens the anal canal but has no effect on the anorectal angle as originally thought. There is growing evidence that anterior pelvic floor repair and sphincteroplasty or total pelvic floor repair may be of more benefit to patients with neurogenic incontinence.

Anal encirclement

Some incontinent patients are unsuitable for sphincter repair or post-anal repair because there is either insufficient functional muscle for direct repair or neurogenic damage is profound. Anal encirclement procedures have been described to treat these patients and patients who have remained incontinent after other procedures. The simplest operation is the Theirsch wire. The anal canal and anal sphincters are supported by silver or steel wire passed around the anus in the subcutaneous plane. This approach to incontinence was popularized by Gabriel.[81]

The traditional Theirsch operation is associated with many complications, including sepsis, sloughing of the overlying skin with extrusion of the wire and faecal impaction. More recent modifications of the technique have used Dacron impregnated silastic sheet to encircle the anal canal,[82] which has the advantage of being somewhat elastic. An incision is made on each side of the anus and deepened into each ischiorectal fossa. A tunnel is created around the anal canal and the silastic strip is passed round the anus and sutured together. A finger in the anal canal is used as a guide to how tight the sling is created. The reported results of this operation are disappointing. Infection, requiring removal of the sling often occurs despite the use of prophylactic antibiotics and a mechanical bowel preparation. Stricker *et al.*[83] treated 14 patients by silastic sling encirclement. Only seven patients improved and five patients required removal of the sling because of septic complications. We have not used this procedure to treat faecal incontinence.

The major problem with the Theirsch operation is that the sling is adynamic and merely narrows a patulous anal canal. An artificial sphincter, made of silastic, has been used to treat urinary incontinence and extensive experience has shown this to be safe and effective. Recently, a modification of this artificial sphincter has been used to treat patients with faecal incontinence.[84] We have been participating in a three centre evaluation of this modified AMS 800 (Americal Medical Systems) urinary sphincter (Fig. 11.1). The sphincter consists of a pressure-regulating balloon, a control pump and an occlusive cuff, all connected together by silastic tubing. The patient is able to regulate opening and closing of the sphincter by manipulating the control pump, implanted in the scrotum or labia majora. When the pump is squeezed, fluid flows from the pressurized cuff into the balloon, deflating the cuff and allowing evacuation of the rectum. The balloon

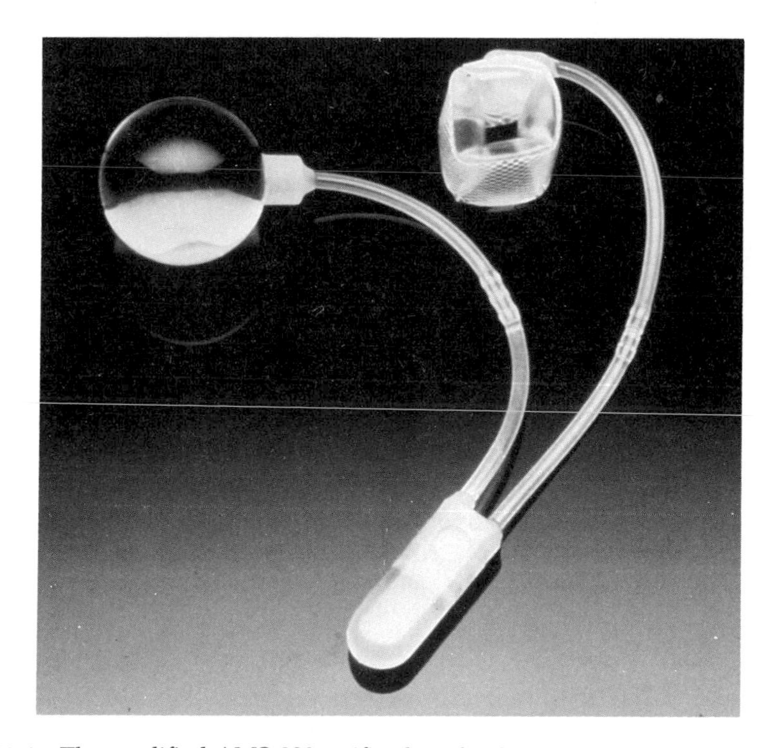

Fig. 11.1 The modified AMS 800 artificial anal sphincter, showing the cuff that is placed round the anal canal, the regulating balloon and the control pump, which are placed in the scrotum or labia majora.

then repressurizes the cuff over the next few minutes and continence is restored.

Implantation of the device is similar to the silastic sling operation. The patient receives complete bowel preparation with oral lavage solution and antibiotics. A defunctioning colostomy is created if not already present. The ischiorectal fossae are entered bilaterally and a tunnel is created around the anal canal, outside the anal sphincter. The tunnel passes deep into the anococcygeal raphe posteriorly and transverse perinei muscle anteriorly. This ensures that the cuff lies in the correct position round the anal canal, deep enough to prevent erosion through the skin. The connecting tube from the cuff is tunnelled subcutaneously into the suprapubic region and the pump is placed in a pocket in the scrotum or labia majora. The pressure-regulating balloon is placed into the space of Retzius through a small Pfannensteil incision. Connections between the pump and the regulating balloon and cuff are made subcutaneously and all wounds are carefully closed in layers. The system is left deactivated for 6 weeks after implantation. The first activation is performed in the anorectal physiology laboratory to document changes in anal canal pressures. The colostomy is closed once the patient is comfortable with manipulating the device.

Christiansen and Lorentzen[85] reported five patients treated with this device. The sphincter was removed from one patient because of infection. Three of the other patients obtained a good result and were continent to solid and liquid stool. The fourth patient had occasional leakage of liquid stool but was continent to solid stool. We have implanted ten artificial anal sphincters in nine incontinent patients. The indication for surgery and results of the operation are shown in Table 11.7. These results are very encouraging, although a more extensive experience with longer follow-up is required before this device can be recommended for standard use.

Muscle transfer

Muscle transfer techniques have been applied to patients with severe loss of sphincter muscle, either congenital in origin or arising from perineal trauma or overly enthusiastic fistula surgery. The inferior half of the gluteus maximus muscles have been transferred to surround the anal canal, carefully preserving the nerve and blood supply to the mobilized muscle. Reasonable results have been reported in case reports of this technique.[86–89] Another approach utilizes the gracilis muscle to encircle the anus. This muscle is the most medial strap muscle of the thigh and the neurovascular supply enters proximally, allowing full mobilization of the muscle after division of its distal insertion. The technique was first described by Pickrell *et al.*[90] and has been described in detail elsewhere.[90] The muscle is mobilized through three separate thigh incisions and passed round the anus in an extra-sphincteric tunnel. The tendon is sutured to the fascia of the opposite ischial tuberosity. The reported results of this operation vary. Corman[92] described 14 patients followed for over 5 years, seven of whom obtained an excellent result and only three patients

Table 11.7 Results of artificial anal sphincter implants

Implant number	PT		Aetiology of incontinence	Stoma	Complications	Months* of function	Continence		
	Gender	Age (years)					Gas	Liquid	Solid
1	Female	31	Birthing injury	Prior	None	25	+	+	+
2	Female	42	Birthing injury	None	Infection – removed (redone – see #8)	0			
3	Female	29	Spina bifida	Prior	None	8	+	+	+
4	Male	29	Trauma	Prior	None	10	+	+	+
5	Male	35	Imperforate anus	Concomitant	None	6	–	+	+
6	Male	15	Trauma	Prior	None	9	+	+	+
7	Male	24	Trauma	Prior	None	6	+	+	+
8	Female	42	Birthing injury	Prior	None	5	+	+	+
9	Male	52	S/P multiple laminectomies	Concomitant	Current	Pending			
10	Male	34	S/P spinal cord tumor	Concomitant	Current	Pending			

*Calculated from time of activation of device after takedown of colostomy. S/P status/position.

went on to require a colostomy for persisting incontinence. Leguit *et al.*[93] noted improvement in nine of ten patients treated by graciloplasty, with six patients achieving continence of stool. However, Yoshioka and Keighley[94] reported very poor results in six patients, none of whom improved and all patients eventually required a colostomy.

The physiological effect of graciloplasty remains uncertain. It is likely that the transposed muscle acts as a dynamic Theirsch repair, which dilates when the patient squats. Manometric studies have demonstrated raised sphincter pressures in some patients following graciloplasty, although some patients with a good clinical result had very low sphincter pressures and success appeared to be a consequence of scarring and narrowing of the anal canal.[93]

One of the disadvantages of muscle transposition is that activity of the transposed muscle depends on voluntary contraction, which can only be sustained for a short time. A novel approach to overcoming this was demonstrated by Baeten *et al.*,[95] who inserted an implantable nerve stimulator adjacent to the nerve supplying the transposed gracilis muscle. Activation of the stimulator resulted in complete continence, with an objective and sustained rise in anal canal pressures. Experimental work has shown that chronic stimulation of the nerve supplying a muscle results in transformation of the muscle from a fast-twitch easily-fatiguable muscle, into a slow-twitch muscle that can sustain continuous contraction for prolonged periods of time.[96] This has obvious application for the construction of a neosphincter using a transposed muscle. A number of incontinent patients have been treated by gracilis transposition and insertion of an implantable electrical stimulator. The early results are very encouraging. Williams *et al.*[97] reported improvement in continence in five of six patients treated, and Baeten *et al.*[98] reported that seven of nine patients became completely continent after this operation. The technique has also been applied to patients undergoing anal excision for carcinoma or inflammatory bowel disease, where a new anus is created using the stimulated gracilis as a neosphincter.[99] This technique is still experimental and we are involved in research to determine the optimum method of chronic nerve stimulation and neosphincter construction.

Correction of the underlying disease

Faecal incontinence is a common accompaniment of rectal prolapse. Following repair of the rectal prolapse, incontinence will improve in about 60 per cent of these patients. Therefore, our approach to the incontinent patient undergoing an abdominal repair of rectal prolapse, is to repair the prolapse first by suture rectopexy and rectosigmoid resection. If incontinence still remains a problem for the patient, a postanal repair or total pelvic floor repair is performed at a subsequent operation. Prolapse patients with anal incontinence treated by perineal rectosigmoidectomy undergo a concomitant levatoroplasty either anteriorly, posteriorly or both anteriorly and posteriorly.

Palliative surgery

Colostomy should be considered for patients with gross faecal incontinence who are unsuitable for any of the preceding procedures, or who remain incontinent despite attempts at surgical correction. The controllable discharge of faeces and flatus into a colostomy appliance will considerably improve the patient's quality of life. We prefer a divided end sigmoid colostomy to a loop colostomy for permanent faecal diversion, as an appliance is easier to fit and the distal end is totally defunctioned.

Conclusions

Faecal incontinence is a distressing affliction for the patient, and severely affects their life style. Patients need full clinical assessment to determine the aetiology of the incontinence and physiological testing is a useful adjunct. A variety of surgical and non-surgical treatments are available, which should be tailored to the individual patient depending on the aetiology of the incontinence. A significant improvement in continence can be achieved in most patients. Colostomy should be considered in patients who remain severely incontinent. New approaches such as an artificial anal sphincter and stimulated transposed muscle techniques are promising developments that will benefit patients unsuitable for more conventional techniques.

References

1 Mandelstam D A. Faecal incontinence: social and economic factors. In Henry M M, Swash M (Eds), *Coloproctology and the Pelvic Floor. Pathophysiology and Management*. London: Butterworths, 1985, pp. 217–22.
2 Thomas T M, Ruff C, Karran O, *et al.*. Study of the prevalence and management of patients with faecal incontinence in old people's homes. *Community Med.* 1987; **9**: 232–7.
3 Rao S C, Read N W, Davidson P A, *et al.*. Anorectal sensitivity and response to rectal distension in patients with ulcerative colitis. *Gastroenterology*. 1987; **93**: 1270–5.
4 Deverode G, Vobecky S, Massé G, *et al.*. Ischaemic fecal incontinence and rectal angina. *Gastroenterology*. 1982; **83**: 970–80.
5 Lane R H S, Parks A G. Function of the anal sphincter following colo-anal anastomosis. *Br J Surg*. 1977; **64**: 596–9.
6 Wexner S D, Jensen L L, Rothenberger D A, *et al.*. Long-term functional analysis of the ileoanal resevoir. *Dis Colon Rectum*. 1989; **32**: 275–81.
7 Duthie H L, Gairns F W. Sensory nerve-endings and sensation in the anal region of man. *Br J Surg*. 1960; **47**: 585–95.
8 Schuster M M, Hendrick T R, Mendeloff A I. The internal anal sphincter response: manometric studies on its normal physiology, neural pathways and alteration in bowel disorders. *J Clin Invest*. 1963; **42**: 196–207.

9 Duthie H L, Bennett R C. The relation of sensation in the anal canal to the functional anal sphincter; a possible factor in anal continence. *Gut*. 1963; **4**: 179–82.
10 Read N W, Abouzekry L, Read M, *et al*.. Anorectal function in elderly patients with fecal impaction. *Gastroenterology*. 1985; **89**: 959–66.
11 Read N W, Abouzekry L. Why do patients with faecal impaction have faecal incontinence. *Gut*. 1986; **27**: 283–7.
12 Wald A, Tunuguntla A K. Anorectal sensorimotor dysfunction in fecal incontinence and diabetes mellitus: modification with biofeedback therapy. *N Eng J Med*. 1984: **310**: 1282–7.
13 Miller R, Bartolo D C C, Cervero F, Mortensen N J McC. Differences in anal sensation in continent and incontinent patients with perineal descent. *Int J Colorectal Dis*. 1989; **4**: 45–9.
14 Miller R, Bartolo D C C, Cervero F, Mortensen N J McC. Anorectal temperature sensation. A comparison of normal and incontinent patients. *Br J Surg*. 1987; **74**: 511–15.
15 Hardcastle J D, Parks A G. A study of anal incontinence and some principles of surgical treatment. *Proc R Soc Med*. 1970; **63** (Suppl): 116–18.
16 Parks A G. Anorectal incontinence. *Proc R Soc Med*. 1975; **68**, 681–90.
17 Womack N R, Morrison J F B, Williams N S. The role of pelvic floor denervation in faecal incontinence. *Br J Surg*. 1986; **73**: 404–7.
18 Bartolo D C, Miller R, Mortensen N J McC. Sphincter mechanism of anorectal continence during Valsalva manoeuvers. *Coloproctology*. 1987; **9**: 103–6.
19 Duthie H L, Watts J M. Contribution of the external anal sphincter to the pressure zone in the anal canal. *Gut*. 1965; **6**: 64–8.
20 Frenckner B, von Euler C. Influence of pudendal block on the anal sphincters. *Gut*. 1975; **16**: 482–9.
21 Schuster M M. Motor action of rectum and anal sphincters in continence and defecation. In Code C F, Heidel W (Eds), *Handbook of Physiology*. Washington DC: American Physiological Society, 1968, pp. 2121–40.
22 Milligan E T C, Morgan C N. Surgical anatomy of the anal canal with special reference to anorectal fistulae. *Lancet*. 1934; **ii**: 1150–6.
23 Bannister J, Gibbons C, Read N W. Preservation of faecal continence during rises in intra-abdominal pressure: is there a role for the flap valve. *Gut*. 1987; **28**: 1242–5.
24 Henry M M, Parks A G, Swash M. The pelvic floor musculature in the descending perineum syndrome. *Br J Surg*. 1982; **69**: 470–2.
25 Neill M E, Parks A G, Swash M. Physiological studies of the pelvic floor musculature in idiopathic faecal incontinence and rectal prolapse. *Br J Surg*. 1981; **68**: 531–6.
26 Roe A M, Bartolo D C C, Mortensen N J McC. New method of assessment of anal sensation in various anorectal disorders. *Br J Surg*. 1986; **73**: 310–12.
27 Rogers J, Henry M M, Misiewicz J J. Combined sensory and motor deficit in primary neuropathic faecal incontinence. *Gut*. 1986; **29**: 5–9.
28 Parks A G, Swash M, Urich H. Sphincter denervation in anorectal incontinence and rectal prolapse. *Gut*. 1977; **18**: 656–67.
29 Bartolo D C C, Jarratt J A, Read N W. The use of conventional EMG to assess external sphincter neuropathy in man. *J Neurol Neurosurg Psychiatry*. 1983; **46**: 1115–8.
30 Neill M E, Swash M. Increased motor unit fibre density in the external anal sphincter muscle in anorectal incontinence; a single fibre EMG study. *J Neurol Neurosurg Psychiatry*. 1980; **43**: 343–7.
31 Kiff E S, Swash M. Normal proximal and delayed distal conduction in the

pudendal nerves of patients with idiopathic (neurogenic) anorectal incontinence. *J Neurol Neurosurg Psychiatry*. 1984; **47**: 820–3.

32 Snooks S J, Swash M, Setchell M, Henry M M. Injury to innervation of pelvic sphincter musculature in childbirth. *Lancet*. 1984; **ii**: 546–50.

33 Snooks S J, Swash M, Henry M M, Setchell M. Risk factors in childbirth causing damage to the pelvic floor innervation. *Br J Surg*. 1985; **72** (Suppl): S15-S17.

34 Lubowski D Z, Swash M, Nicholls R J, Henry M M. Increase in pudendal nerve terminal motor latency with defaecation straining. *Br J Surg*. 1988; **75**: 1095–7.

35 Jones P N, Lubowski D Z, Swash M, Henry M M. Relation between perineal descent and pudendal nerve damage in idiopathic faecal incontinence. *Int J Colorectal Dis*. 1987; **2**: 93–5.

36 Williams J G, Wong W D, Jensen L, *et al.*. Incontinence and rectal prolapse: a prospective manometric study. *Dis Colon Rectum*. 1991; **34**: 209–16.

37 Hiltunen K M, Matikainen M, Auvunen O, Hietanen P. Clinical and manometric evaluation of anal sphincter function in patients with rectal prolapse. *Am J Surg*. 1986; **151**: 489–92.

38 Metcalf A M, Loening-Baucke V. Anorectal function and defecation dynamics in patients with rectal prolapse. *Am J Surg*. 1988; **155**: 206–10.

39 Keighley M R B, Makuria T, Alexander-Williams J, Arabi Y. Clinical and manometric evaluation of rectal prolapse and incontinence. *Br J Surg*. 1980; **67**: 54–6.

40 Matheson D M, Keighley M R B. Manometric evaluation of rectal prolapse and faecal incontinence. *Gut*. 1981; **22**: 126–9.

41 Ihre T, Seligson U. Intussusception of the rectum – Internal procidentia: Treatment and results in 90 patients. *Dis Colon Rectum*. 1975; **18**: 391–6.

42 Holmström B, Brodén G, Dolk A, Frenckner B. Increased anal resting pressure following the Ripstein operation. A contribution to continence. *Dis Colon Rectum*. 1986; **29**: 485–7.

43 Snooks S J, Henry M M, Swash M. Anorectal incontinence and rectal prolapse: differential assessment of the innervation to puborectalis and external anal sphincter muscles. *Gut*. 1985; **26**: 470–6.

44 Brodén G, Dolk A, Holmström B. Recovery of the internal anal sphincter following rectopexy: a possible explanation for continence improvement. *Int J Colorectal Dis*. 1988; **3**: 23–8.

45 Bartolo D C C, Duthie G S. The physiological evaluation of operative repair for incontinence and prolapse. In Bock L, Whelan J (Eds), *Neurobiology of Incontinence*. Chichister: John Wiley and Sons, 1990, pp. 223–450.

46 Yoshioka K, Hyland G, Keighley M R B. Anorectal function after abdominal rectopexy: parameters of predictive value in identifying return of continence. *Br J Surg*. 1989; **76**: 64–8.

47 Laurberg S, Swash M, Henry M M. Delayed external sphincter repair for obstetric tear. *Br J Surg*. 1988; **75**: 786–8.

48 Miller R, Bartolo D C C, Locke-Edmunds J C, Mortensen N J McC. Prospective study of conservative and operative treatment for faecal incontinence. *Br J Surg*. 1988; **75**: 101–5.

49 Read N W, Harford W V, Schmulen A C, *et al.*. A clinical study of patients with fecal incontinence and diarrhea. *Gastroenterology*. 1979; **76**: 747–56.

50 Engel B T, Nikoomanesh P, Schuster M M. Operant conditioning of rectosphincteric responses in the treatment of fecal incontinence. *N Engl J Med*. 1974; **290**: 646–9.

51 Cerulli M A, Nikoomanesh P, Schuster M M. Progress in biofeedback conditioning for fecal incontinence. *Gastroenterology*. 1979; **76**: 742–6.
52 Goldenberg D A, Hodges K, Hersh T, Jinich H. Biofeedback therapy for fecal incontinence. *Am J Gastroenterol*. 1980; **74**: 342–5.
53 Wald A. Biofeedback therapy for fecal incontinence. *Ann Intern Med*. 1981; **95**: 146–9.
54 Whitehead W E, Burgio K L, Engel B T. Biofeedback treatment of fecal incontinence in geriatric patients. *J Am Geriatric Soc*. 1985; **33**: 320–4.
55 MacLeod J H. Management of anal incontinence by biofeedback. *Gastroenterology*. 1987; **93**: 291–4.
56 Riboli E B, Francio M, Pitto G, *et al*.. Biofeedback conditioning for fecal incontinence. *Arch Phys Med Rehabil*. 1988; **69**: 29–31.
57 Jensen L L, Lowry A C. Biofeedback: a viable treatment option for anal incontinence. *Dis Colon Rectum*. 1991; **34**: 6.
58 Loening-Baucke V, Desch L, Wolraich M. Biofeedback training in patients with meningomyelocele and fecal incontinence. *Dev Med Child Neurol*. 1988; **30**: 781–90.
59 Wald A. Biofeedback for neurogenic fecal incontinence: Rectal sensation is a determinant of outcome. *J Pediatr Gastroenterol Nutr*. 1983; **2**: 302–6.
60 Rothenberger DA. Anal incontinence. In Corman M L (Ed), *Current Surgical Therapy (vol 3)*. Philadelphia: BC Decker, 1989, pp. 185–94.
61 Castro A F, Pittman R E. Repair of the incontinent sphincter. *Dis Colon Rectum*. 1978; **21**: 183–7.
62 Keighley M R B, Fielding J W L. Management of faecal incontinence and results of surgical treatment. *Br J Surg*. 1983; **70**: 463–8.
63 Browning G G P, Motson R W. Anal sphincter injury. Management and results of Parks sphincter repair. *Ann Surg*. 1984; **199**: 351–7.
64 Fang D T, Nivatvongs S, Vermeulen F D, *et al*.. Overlapping sphincteroplasty for acquired anal incontinence. *Dis Colon Rectum*. 1984; **27**: 720–2.
65 Pezim M E, Spencer R J, Stanhope C R, *et al*.. Sphincter repair for fecal incontinence after obstetrical or iatrogenic injury. *Dis Colon Rectum*. 1987; **30**: 521–5.
66 Yoshioka K, Keighley M R B. Sphincter repair for fecal incontinence. *Dis Colon Rectum*. 1989; **32**: 39–42.
67 Miller R, Orrom W J, Cornes H, *et al*.. Anterior sphincter plication and levatoroplasty in the treatment of faecal incontinence. *Br J Surg*. 1989; **76**: 1058–60.
68 Wexner S D, Marchetti F, Jagelman D G. The role of sphincteroplasty for fecal incontinence reevaluated: a prospective physiologic and functional review. *Dis Colon Rectum*. 1991; **34**: 22–30.
69 Ctercteko G C, Fazio V W, Jagelman D G, *et al*.. Anal sphincter repair: a report of 60 cases and review of the literature. *Aust NZ J Surg*. 1988; **58**: 703–10.
70 Motson R W. Sphincter injuries: indications for and results of sphincter repair. *Br J Surg*. 1985; **72** (Suppl): S19–21.
71 Jacobs P P M, Scheuer M, Kuijpers J H C, Vingerhoets M H. Obstetric fecal incontinence. Role of pelvic floor denervation and results of delayed sphincter repair. *Dis Colon Rectum*. 1990; **33**: 494–7.
72 Roberts P L, Coller J A, Schoetz D J, Veidenheimer M C. Manometric assessment of patients with obstetric injuries and fecal incontinence *Dis Colon Rectum*. 1990; **33**: 16–20.
73 Parks A G. Post-anal pelvic floor repair (and the treatment of faecal incontinence). In Rob C, Smith R (Eds), *Operative Surgery. Colon Rectum and Anus* (3rd edn). London: Butterworths, 1977, pp. 249–54.

74 Browning G G P, Parks A G. Postanal repair for neuropathic faecal incontinence: correlation of clinical result and anal canal pressures. *Br J Surg*. 1983; **70**: 101–4.
75 Henry M M, Simson J N L. Results of postanal repair: a retrospective study. *Br J Surg*. 1985; **72** (Suppl): S17–S19.
76 Yoshioka K, Hyland G, Keighley M R B. Clinical and physiological evaluation of postanal repair. *Gut*. 1987; **28**, A1362.
77 Womack N R, Morrison J F B, Williams N S. Prospective study of the effects of post-anal repair in neurogenic faecal incontinece. *Br J Surg*. 1988; **75**: 48–52.
78 Yoshioka K, Hyland G, Keighley M R B. Physiological changes after postanal repair and parameters predicting outcome. *Br J Surg*. 1988; **75**: 1220–4.
79 Scheuer M, Kuijpers H C, Jacobs P P. Post-anal repair restores anatomy rather than function. *Dis Colon Rectum*. 1989; **32**: 960–3.
80 Rainey J B, Donaldson D R, Thomson J P S. Postanal repair: which patients derive most benefit? *J R Coll Surg Edinburgh*. 1990; **35**: 101–5.
81 Gabriel W B. Theirsch's operation for anal incontinence. *Proc R Soc Med*. 1948; **46**: 467–8.
82 Labow S, Rubin R J, Hoexter B, Salvati E P. Perineal repair of rectal procidentia with an elastic fabric sling. *Dis Colon Rectum*. 1980; **23**: 467–9.
83 Stricker J W, Schoetz D J, Clooer J A, Veidenheimer M C. Surgical correction of anal incontinence. *Dis Colon Rectum*. 1988; **31**: 533–40.
84 Christiansen J, Lorentzen M. Implantation of artificial sphincter for anal incontinence. *Lancet*. 1987; **ii**: 244–5.
85 Christiansen J, Lorentzen M. Implantation of artificial sphincter for anal incontinence. *Dis Colon Rectum*. 1989; **32**: 432–6.
86 Hentz V R. Construction of a rectal sphincter using the origin of the Gluteus maximus muscle. *Plast Reconstr Surg*. 1982; **70**: 82–5.
87 Skef Z, Radhakrishan J, Reyes H M. Anorectal continence following sphincter reconstruction utilizing the gluteus maximus muscle: a case report. *J Pediatr Surg*. 1983; **18**: 779–81.
88 Iwai N, Kaneda H, Tsuto T, *et al.*. Objective assessment of anorectal function after sphincter reconstruction using the gluteus maximus muscle. *Dis Colon Rectum*. 1985; **28**: 973–7.
89 Pearl R K, Prasad M L, Nelson R L, *et al.*. Bilateral gluteus maximus transposition for anal incontinence. *Dis Colon Rectum*. 1991; **34**: 478–81.
90 Pickrell K L, Broadbent T R, Masters F W, Metzger J T. Construction of a rectal sphincter and restoration of anal continence by transplanting the gracilis muscle: a report of four cases in children. *Ann Surg*. 1952; **135**: 853–62.
91 Corman M L. Anal incontinence. In Corman M L (Ed), *Colon and Rectal Surgery* (2nd edn). Philadelphia: J B Lippincott, 1989, pp. 192–200.
92 Corman M L. Gracilis muscle transposition for anal incontinence: late results. *Br J Surg*. 1985; **72** (Suppl): S21–22.
93 Leguit P, Jr, Van Baal J G, Brummelkamp W H. Gracilis muscle transposition in the treatment of fecal incontinence. *Dis Colon Rectum*. 1985; **28**: 1–4.
94 Yoshioka K, Keighley M R B. Clinical and manometric assessment of gracilis muscle transplant for fecal incontinence. *Dis Colon Rectum*. 1988; **31**: 767–9.
95 Baeten C, Spaans F, Fluks A. An implanted neuromuscular stimulator for fecal continence following previously implanted gracilis muscle. *Dis Colon Rectum*. 1988; **31**: 134–7.
96 Mannion J D, Bitto T, Hammond R L, *et al.*. Histochemical and fatigue characteristics of conditioned canine latissimus dorsi muscle. *Circulation Res*. 1986; **58**: 2.
97 Williams N S, Hallan R I, Koeze T H, *et al.*. Construction of a neoanal

sphincter by transposition of the gracilis muscle and prolonged neuromuscular stimulation for the treatment of faecal incontinence. *Ann R Coll Surg Engl.* 1990; **72**: 108–13.

98 Baeten C, Konsten J, Spaans F, *et al.*. Dynamic graciloplasty for fecal incontinence. *Dis Colon Rectum.* 1991; **34**: 6–7.

99 Cavina E, Seccia M, Evangelista G, *et al.*. Perineal colostomy and electrostimulated gracilis 'neosphincter' after abdomino-perineal resection of the colon and anorectum: a surgical experience and follow-up study in 47 cases. *Int J Colorectal Dis.* 1990; **5**: 6–11.

12

Pyoderma gangrenosum in inflammatory bowel disease

Michael Levitt and Robin K.S. Phillips

Introduction

Pyoderma Gangrenosum (PG) is a painful chronic ulcerating skin disease of unknown cause. Typically it commences as a sterile erythematous papulopustule that evolves rapidly, sometimes coalescing with adjacent lesions, to form an area of frank ulceration. Characteristically the ulcer has an undermined, violaceous edge and a bright outer halo of erythema. The diagnosis is essentially clinical as there are no accepted diagnostic histological features.

In as many as 78 per cent of patients with PG an underlying systemic illness is present,[1] the most prominent being inflammatory bowel disease – both ulcerative colitis and Crohn's disease – which accounts for 13–56 per cent of cases of PG.[1–5] In addition, a variety of rheumatological, haematological and other diseases have been associated with PG (Table 12.1).

In cases where there is no demonstrable underlying systemic disease the clinical pattern of PG is no different. As PG may precede the onset of an associated systemic disorder by many years the number of cases of truly isolated PG is unknown and may have previously been overestimated.

Pyoderma gangrenosum in inflammatory bowel disease

Pyoderma gangrenosum is said to develop at some stage in the course of 0.5–5.0 per cent of patients with ulcerative colitis[1,6–9] and 0.3–1.5 per cent of patients with Crohn's disease.[1,9–11] The intestinal component generally appears first with PG nearly always arising in the next 10 years.[1,4,6] Classically, the activity of the skin lesion is said to parallel that of the intestinal disease with complete excision of diseased intestine producing prompt skin healing.[1,12,13] This classical pattern is not always evident. Pyoderma gangrenosum precedes the onset of inflammatory bowel disease in as many as 14 per cent of cases,[5,6] may run a course independent of the activity of the intestinal disease,[4,7,11] may respond

Table 12.1 Systemic diseases associated with pyoderma gangrenosum (PG)

Gastrointestinal
Ulcerative colitis
Crohn's disease
Diverticular disease
Rheumatological
Sero-negative large joint oligo-arthritis
Sero-negative symmetric small joint polyarthritis
Sero-positive rheumatoid arthritis
Ankylosing spondylitis
Osteoarthritis
Systemic lupus erythematosus
Haematological
Monoclonal gammopathy
Myeloma
Leukaemia
Polycythaemia rubra vera
Primary thrombocythaemia
Other
Chronic active hepatitis
Hyperthyroidism
Metastatic malignancy
Diabetes mellitus
Sarcoidosis

poorly to total resection of diseased intestine[14] and has been reported to develop years after total proctocolectomy for ulcerative colitis.[6,7,15–19] The precise nature of the relationship between PG and inflammatory bowel disease remains unresolved.

Pyoderma gangrenosum in inflammatory bowel disease at St Mark's hospital

From 1954–90, 34 patients were seen at St Mark's Hospital with a clinical diagnosis of PG associated with either ulcerative colitis (22 patients) or Crohn's disease (12 patients). Their clinical details are summarized in Table 12.2.

Age and sex distribution

Although there was a wide age range, three-quarters of the cases were between 20 and 60 years old. In the largest published series of PG 'most' patients were also noted to range from 25–54 years of age.[1] Thornton *et al.*[4] reported an average age of 45 years amongst six patients with PG associated

Table 12.2 Clinical details of 34 patients with pyoderma gangrenosum (PG) in association with ulcerative colitis and Crohn's disease

	Ulcerative colitis	Crohn's disease
Sex distribution (Male:Female)	8:14	6:6
Age in years at onset of PG (median, range)	32,15–77	27.5, 16–56
Number of skin lesions (single:multiple)	6:16	4:8
Sites of skin lesions		
legs (below knee)	26	16
thighs/buttocks	10	3
upper limbs	2	3
chest/back	6	–
head and neck	3	1
peristomal/wound	2	1
History of pathergy*	3	1
Associated illnesses	12 (55%)	11 (92%)
Sero-negative arthritis	5	9
Erythema nodosum	4	5
Iritis/episcleritis	3	6
Oral ulceration	2	3
Pustular rash	3	3
Abscess at distant site	2	5
Extent of intestinal disease at onset of PG	Total colitis 12	Anus only 1
	Left-sided colitis** 8	Anus + colon 6
	Rectum after colectomy + ileo-rectal anastomosis 2	Colon only 2
		Ileum + colon 2
		Mouth only 1
Intestinal disease symptomatically active at onset of PG	11[†] (50%)	9[††] (75%)

*Ulceration developing at the site of previous skin trauma. **Distal to the splenic flexure on radiology. [†]Seven had total colitis, three had left-sided colitis, and one rectal involvement after colectomy and ileorectal anastomosis. [††]Of the three patients with inactive disease at the onset of PG, one had oral ulceration only, one had anal and perianal disease with no involvement of colorectal mucosa and the third had recurrent ileal disease with minimal colonic ulceration after prior right hemicolectomy. Reproduced by permission of the publishers, Butterworth-Heinemann Ltd from Levitt *et al.*, *Br J Surg*. 1991; **78**: 676–8.

with ulcerative colitis; patients with PG unassociated with ulcerative colitis were significantly older (mean 69 years old) in this report. Both groups, however, were considerably older than those seen at St Mark's Hospital (median age for all patients = 32 years).

There was no significant sex preponderance in accordance with previous reports.[1] There was, however, a tendency (p = non-significant) for affected males to be older (16–77 years, median age = 46) than affected females (15–59, median age = 30).

Site of skin lesions

Lesions at multiple sites were noted in a substantial majority of ulcerative colitics (16 patients, 73 per cent) and patients with Crohn's disease (eight patients, 67 per cent). Although multiple lesions are well recognised, this high incidence is noteworthy.

The propensity for PG to occur on the lower limbs – over 50 per cent of all lesions were seen below the knees – is characteristic of this disease.[1,11,20] The 12 per cent incidence of pathergy – the tendency for PG to develop at the site of previous skin trauma – is also in accordance with previously reported rates ranging up to one-third of cases.[1,11,19] Pyoderma gangrenosum was noted to be peristomal or affecting a recent abdominal incision on three occasions, a manifestation of pathergy that is well recognized.[2,16–18]

Associated illnesses

Arthritis

An associated sero-negative arthritis is well described. There are two common patterns: either an asymmetric large joint arthropathy as is often seen in inflammatory bowel disease; or a more diffuse, erosive polyarthropathy. Holt *et al.*[2] noted sero-negative arthritis in six of 15 patients (40 per cent) with PG, whereas Powell *et al.* recorded 24 patients with sero-negative arthritis among 86 patients (28 per cent), the majority of these being associated with inflammatory bowel disease.[1] A further report described joint pain and swelling in 38 per cent of patients with PG associated with ulcerative colitis.[6]

At St Mark's Hospital 14 patients (41 per cent) were noted to have some form of sero-negative arthritis. This was mostly of the large joint variety affecting the knees in particular but in three patients a diffuse polyarthritis was recorded. There was no evidence of erosion or progression in these three patients. Arthritis was more frequently associated with PG in patients with Crohn's disease (nine of 12, 75 per cent) than with ulcerative colitis (five of 22, 23 per cent).

Erythema nodosum

Erythema nodosum was separately diagnosed in nine patients (26 per cent). By contrast, another study diagnosed the two skin conditions in the same patient only once in 42 patients (2 per cent) with ulcerative colitis.[6] Schoetz *et al.*[11] reported eight patients with both Crohn's disease and PG; erythema nodosum was diagnosed in only one. The high frequency with which these two skin lesions were separately diagnosed in patients with inflammatory bowel disease at St Mark's Hospital is difficult to explain. Certainly the early lesions of PG resemble erythema nodosum but in this series the diagnosis depended upon frank ulceration. Perhaps the two processes are more frequently concurrent than has previously been considered.

Pyoderma-like lesions

A pustular rash occurring over the legs, buttocks, trunk, arms or face that did not progress to frank ulceration was seen in six patients. In five of the six it appeared at the same time as PG. Other authors have described a pustular rash,[21] pustular dermatitis,[22] scattered papulopustules[6] and pustular 'eruptions'[23] in association with PG. Both hidradenitis suppurativa and acne conglobata, also characterized by pustular lesions, may be seen in patients with PG.[1] The skin lesion of PG commonly commences as a pustule; in the original description of pyoderma in 1930,[24] Brunsting noted that the skin lesions commenced as crops of small discreet pustules that progressed to ulceration. A number of authors have since commented that these 'pyoderma-like' lesions may be regarded as variants of PG;[9,20,23] they may represent an alternative cutaneous manifestation of the same immunological process thought to underlie PG.

Suppuration at distant sites

An association between the development and persistence of PG in the presence of a focus of distant suppuration was noted, particularly in patients with Crohn's disease (Table 12.3). This association may be entirely coincidental; perianal sepsis in particular is a common manifestation of Crohn's disease. But the absence of any active intestinal disease at the onset of PG in three patients (patients one, three and seven), the persistence of PG in the presence of continued perineal sepsis despite resection of all diseased intestine in two patients (two and four) and prompt healing of PG following drainage of sepsis (surgical or spontaneous) without resection in two patients (three and five) raises the possibility that foci of distant suppuration may precipitate or prolong the course of PG.

In Brunstings original description of PG,[24] four of the five cases described were associated with active ulcerative colitis; the fifth, however, occurred in a patient with a chronic empyema. Holmlund and Wahlby[18] reported a case of peristomal PG occurring after proctocolectomy but in the presence of perineal and peristomal abscesses. Apart from these two brief references there has been little to support an association between PG and distant suppuration. Nevertheless there are potential mechanisms for such an association.

Cross antigenicity between *Escherichia coli* (a bacterium frequently cultured in sepsis of intestinal origin) and cutaneous antigens has been suggested by the successful treatment of previously unresponsive PG in two patients by selective elimination of *E coli* from the gastrointestinal tract.[25] It is therefore possible that an organism such as *E coli* serves as an antigen against which antibodies may be formed that are also directed against cutaneous antigens. Alternatively, other non-microbial components of an abscess cavity might serve as the source of cross antigenicity with skin. Such components may equally reside in the crypt micro-abscesses of inflammatory bowel disease, thereby explaining the association of PG with active intestinal disease.

Although speculative, these observations suggest that consideration be given to the possibility of co-existent, undrained perineal, pelvic or more

Table 12.3 Association of pyoderma gangrenosum (PG) with distant suppuration.

Patient	Underlying inflammatory bowel disease	Clinical summary
1	Ulcerative colitis	Unsuspected, large pelvic abscess found at proctocolectomy performed for PG resistant to medical therapy and symptomatically quiescent left-sided colitis. PG healed promptly after surgery.
2	Ulcerative colitis	Left-sided colitis with recto-vaginal fistula and extensive perianal suppuration at onset of PG. Construction of loop ileostomy complicated by wound infection and perianal abscess formation with persistence of PG. Proctocolectomy complicated by further perineal and para-ileostomy abscess formation and worsening PG. PG remained unhealed until perineal sepsis was controlled 2 months after proctocolectomy.
3	Crohn's disease	Multiple peri-anal and enterocutaneous fistulae associated with ileal and colonic Crohn's disease at onset of PG. Proctocolectomy with ileal resection allowed PG to heal. Recurrence of PG 6 months later while asymptomatic, remaining resistant to medical therapy for 16 months. At that time, spontaneous perineal discharge of an enterocutaneous fistula was associated with prompt resolution of PG. No radiological evidence of Crohn's disease; fistula resolved spontaneously.
4	Crohn's disease	Distal colonic disease only. Onset of PG associated with multiple perianal fistulae and discharge of pus per rectum. Incomplete response to both medical therapy and colectomy with ileostomy. Rectal excision complicated by further perineal sepsis; PG on leg healed but new lesion appeared around stoma remaining unhealed at that site until perineal suppuration was controlled 4 months after rectal excision.
5	Crohn's disease	Severe anorectal disease. Groin and peri-anal (Crohn's) ulceration with pus discharging per rectum at onset of PG, remaining resistant to medical therapy for 3 months. Prompt healing of PG followed drainage of a large peri-rectal abscess.
6	Crohn's disease	Flare-up of known total colitis associated with rectovaginal and peri-anal fistulae at onset of PG. PG healed within 2 weeks of medical therapy. Presented 4 weeks later with purulent peritonitis secondary to a ruptured pericolic abscess.

7	Crohn's disease	Colonic disease associated with recurrent PG treated by colectomy and ileorectal anastomosis. Further exacerbation of PG 4 months before a rectovaginal fistula developed. Small bowel found to be entirely normal at laparotomy for concurrent small bowel obstruction. PG then healed but reappeared 12 months later at a different site in association with a dental abscess and septic arthritis but no evidence of active intestinal disease.

Reproduced by permission of the publishers, Butterworth-Heinemann Ltd, from Levitt *et al.*, *Br J Surg*. 1991: **78**: 676–8.

distant abscess formation when PG is resistant to medical therapy or persists after resection of all diseased intestine.

The relation of pyoderma gangrenosum to the extent of intestinal disease

Colitis was total in 12 of the 22 patients with PG associated with ulcerative colitis, was distal to the splenic flexure in eight and was confined to the rectum (after previous colectomy and ileorectal anastomosis) in two. In a study of 220 patients with ulcerative proctitis and mild ulcerative colitis,[8] three of 159 patients with disease confined to the rectum and distal sigmoid colon (1.9 per cent) had PG; this was no different to the overall incidence of PG in ulcerative colitics at that institution. Johnson and Wilson[7] were also unable to identify any parallel between disease extent and severity of PG amongst 415 patients with ulcerative colitis. In ulcerative colitis the extent of diseased intestine appears to bear no relation to the likelihood of developing PG.

On the other hand, Farmer *et al.*[10], noting that colonic disease was present in all nine patients with PG amongst 615 with Crohn's disease, concluded that PG occurring in association with Crohn's disease did so only in the presence of colonic involvement. As a corollary, van Patter *et al.*[26] found only one case of PG amongst 600 patients with isolated small bowel Crohn's disease. At St Mark's Hospital ten of the 12 patients with PG associated with Crohn's disease had colonic involvement, nine of whom had active intestinal disease at the onset of PG. In the tenth patient, Crohn's disease was principally ileal with only minimal colonic ulceration. These observations support the notion that colonic disease is an important if not universal accompaniment of PG in inflammatory bowel disease and draws attention to the colon as a possible source of the antigen or common immunological mediator thought to be responsible for the development of skin lesions.[1] The identification of epidermolytic proteases from the stools of patients with PG,[27] although of uncertain significance, also focuses attention on the colon in the pathogenesis of PG. However, as PG may develop long after excision of all colorectal mucosa[6,7,15–19] the situation remains unclear.

Treatment

The outcome of treatment of PG is summarized in Table 12.4. The generally good response of PG to medical treatment with complete healing in two-thirds of all cases is consistent with numerous other reports.[5,6,14,17] Oral corticosteroids are the mainstay of medical therapy; they were employed in all 16 patients with ulcerative colitis and five of the eight patients with Crohn's disease whose skin lesions healed on medical therapy alone.

Table 12.4 Response to treatment of pyoderma gangrenosum (PG)

	Ulcerative colitis	Crohn's disease
PG healed with medical therapy	16 (67%)	8 (67%)
Intestinal resection when PG present	9*	6**
Response to resection		
Prompt healing	3	3
Healing only with additional therapy	3†	1††
Unhealed or very slow healing	3	2
Recurrent episodes of PG	8 (36%)	5 (42%)

*Nine resections in eight patients. **Six resections in five patients. †Steroids required in two, split skin graft in one. ††Steroids required.

The relation of pyoderma gangrenosum to intestinal disease activity

Ulcerative colitis

There was no obvious correlation between disease activity and the onset or persistence of PG in patients with ulcerative colitis. At the onset of PG, ulcerative colitis was clinically active in only half the patients. Of the nine colonic resections for ulcerative colitis performed when PG was present, skin lesions healed promptly in only three cases, with additional therapy in three cases and very slowly or not at all in three cases. Even when all colorectal mucosa was resected (seven cases) PG healed promptly in only two. Conversely, healing in one case was prompt despite macroscopic disease being left unresected.

Equally, the response to surgical resection was unrelated to colitis activity at the time of resection: when active at the time of colectomy (five cases) healing was prompt in one and achieved only with the help of other therapy in two; in the four cases in whom colitis was inactive at the time of resection healing was prompt in two and achieved only with the help of further therapy in one.

Others have also reported an apparent absence of association between PG and intestinal disease activity in ulcerative colitis.[4,7,28] In particular, Thornton *et al.*[4] reported 14 patients with PG associated with ulcerative colitis in all of whom the intestinal disease was inactive at the onset of skin lesions. Indeed PG is well known to precede the onset of symptomatic ulcerative colitis[1,5,6] or to develop years after total proctocolectomy.[6,7,15–19] Whereas there are undeniably cases in whom the intestinal and skin components run parallel courses and in whom unhealed PG responds promptly to complete resection of diseased intestine, it is not possible confidently to predict cure of the skin lesion by intestinal resection, either on the basis of disease activity at the time of surgery or completeness of resection.

Crohn's disease

There was a much stronger association between disease activity at the onset of PG in patients with Crohn's disease, especially where there was involvement of the large bowel. However, Schoetz *et al.*[11] reported eight patients with PG in association with Crohn's disease in three of whom the intestinal component was inactive at the onset of the skin lesion and Finkel and Janowitz[28] reported the development of PG in one patient with Crohn's disease with asymptomatic intestinal disease.

Six intestinal resections were undertaken in patients with Crohn's disease at a time when PG was present, skin lesions healing promptly in three, only with additional therapy in one and slowly or not at all in two. Resection was undertaken at a time when Crohn's disease was active in all cases and in only one was macroscopic Crohn's disease left unresected – skin healing was protracted in this case. Nevertheless, even when total excision of actively involved Crohn's disease was performed (five resections), healing of the skin lesion was prompt in only three.

The weak association between PG and intestinal disease activity in both Crohn's disease and ulcerative colitis demonstrated at St Mark's Hospital is in direct contrast to the observations of other authors.[2,6,9,13] In particular, the inconsistent response of the skin lesion to resection of all diseased intestine is contrary to previous reports of rapid and complete resolution of PG following resection of all diseased colonic and rectal mucosa.[1,6,14,21] Furthermore, these findings do not support the observations of Talansky *et al.*[14] that healing of PG after intestinal resection is slow when the intestinal disease is inactive at the time of surgery. The response of PG to intestinal resection is unpredictable.

Implications for pathogenesis

Pyoderma gangrenosum is thought to be an immunologically-mediated disease[1,9,29–31] directed against dermal blood vessels. This is based on the demonstration of immune complex deposition in the walls of dermal blood vessels in the majority of patients tested,[1] and is supported indirectly by the association of PG with other immunologically-mediated diseases (e.g., inflammatory bowel disease, arthritis) or diseases associated with an immune deficit (e.g., various haemotological disorders). What is more, patients with PG have been shown to exhibit a number of sub-clinical abnormalities of immune reactivity, such as altered delayed hypersensitivity, abnormal polymorphonuclear leukocyte function and reduced complement levels.[2,20,31] There is, however, no consistent or unifying thread to indicate a specific immune defect.

It is possible that PG is just one of a number of separate though frequently concurrent immunological processes including inflammatory bowel disease, erythema nodosum, arthritis and others, as has been suggested by Greenstein and co-workers.[9] Such an hypothesis has already been applied to the frequently concurrent immunological processes seen in Graves' disease.[32] Alternatively, PG may comprise a number of sub-groups each with a different mechanism of pathogenesis. In some patients a common immunological mediator, possibly residing in intestinal microabscesses, may be active; in others PG may more closely resemble a true autoimmune process. For the present, the pathogenesis of PG in inflammatory bowel disease remains unknown.

References

1 Powell F C, Schroeter A L, Su W P D, Perry H O. Pyoderma gangrenosum: a review of 86 patients. *Q J Med*. 1985; **217**: 173–86.
2 Holt P J A, Davies M G, Saunders K C, Nuki G. Pyoderma gangrenosum. Clinical and laboratory findings in 15 patients with special reference to polyarthritis. *Medicine* (Baltimore). 1980; **59** (2): 114–33.
3 Hickman J G, Lazarus G S. Pyoderma Gangrenosum: A re-appraisal of associated systemic disease. *Br J Dermatol*. 1980; **102**: 235–6.
4 Thorton J R, Teague R H, Low-Beer T S, Read A E. Pyoderma gangrenosum and ulcerative colitis. *Gut*. 1980; **21**: 247–8.
5 Perry H O. Pyoderma gangrenosum. *South Med J*. 1969; **62**: 899–908.
6 Mir-Madjlessi S H, Taylor J S, Farmer R G. Clinical course and evolution of erythema nodosum and pyoderma gangrenosum in chronic ulcerative colitis: a study of 42 patients. *Am J Gastroenterol*. 1985; **80**: 615–20.
7 Johnson M L, Wilson H T H. Skin lesions in ulcerative colitis. *Gut*. 1969; **10**: 255–63.
8 Sparberg M, Fenessey J, Kirsner J B. Ulcerative proctitis and mild ulcerative colitis: a study of 220 patients. *Medicine* (Baltimore). 1966; **45**: 391–412.
9 Greenstein A J, Janowitz H D, Sachar D B. The extra-intestinal complications of Crohn's disease and ulcerative colitis: a study of 700 patients. *Medicine* (Baltimore). 1976; **55**: 401–12.
10 Farmer R G, Hawk W A, Turnbull R B, Jr. Clinical patterns in Crohn's disease: a statistical study of 615 cases. *Gastroenterology*. 1975; **68**: 627–35.

11 Schoetz D J, Jr, Coller J A, Veidenheimer M C. Pyoderma gangrenosum and Crohn's disease. Eight cases and a review of the literature. *Dis Colon Rectum.* 1983; **26**: 155–8.
12 Goligher J C, de Dombal F T, Watts J, Watkinson G. Ulcerative colitis. London: Ballière Tindall, Cassell, 1968, p. 136.
13 Smith J N, Winship D H. Complications and extra-intestinal problems in inflammatory bowel disease. *Med Clin North Am.* 1980; **64**: 1161–71.
14 Talansky A L, Meyers S, Greenstein A J, Janowitz H B. Does intestinal resection heal the pyoderma gangrenosum of inflammatory bowel disease? *J Clin Gastroenterol.* 1983; **5**: 207–10.
15 Cook T J, Lorincz A L. Pyoderma gangrenosum appearing 10 years after colectomy and apparent cure of chronic ulcerative colitis. *Arch Dermatol.* 1962; **86**: 105–6.
16 Margoles J S, Wenger J. Stomal ulceration associated with pyoderma gangrenosum and chronic ulcerative colitis. Report of two cases. *Gastroenterology.* 1961; **41**: 594–8.
17 McGarity W C, Robertson D B, McKeown P P, *et al.*. Pyoderma gangrenosum at the parastomal site in patients with Crohn's disease. *Arch Surg.* 1984; **119**: 1186–8.
18 Holmlund D E W, Wahlby L. Pyoderma gangrenosum after colectomy for inflammatory bowel disease. *Acta Chir Scand.* 1987; **153**: 73–4.
19 Cox N H, Peebles-Brown D A, MacKie R M. Pyoderma gangrenosum occurring ten years after proctocolectomy for ulcerative colitis. *Br J Hosp Med.* 1986; **36**: 363.
20 Basler R S W. Ulcerative colitis and the skin. *Med Clin North Am.* 1980; **64**: 941–54.
21 Hibberd A D, Cuthbertson A M. Systemic manifestations of chronic ulcerative colitis refractory to colectomy. *Aust NZ J Surg.* 1980; **50**: 44–6.
22 Nugent F W, Rudolf N E. Extra-colonic manifestations of chronic ulcerative colitis. *Med Clin North Am.* 1966; **50**: 529–41.
23 O'Loughlin S, Perry H O. A diffuse pustular eruption associated with ulcerative colitis. *Arch Dermatol.* 1978; **114**: 1061–2.
24 Brunsting L A, Goeckerman W H, O'Leary P A. Pyoderma (echthyma) gangrenosum. *Arch Dermatol Syphilis* (Chicago). 1930; **22**: 655–80.
25 Driessen L H H M, van Saene H K F. A novel treatment of pyoderma gangrenosum by intestinal decontamination. *Br J Dermatol.* 1983; **108**: 108 (Abstr).
26 Van Patter W N, Bargen J A, Dockerty M B, *et al.*. Regional enteritis. *Gastroenterology.* 1954; **26**: 347–450.
27 Stoughton R B. Enzymatic cytolysis of epithelium by filtrates from patients with ulcerative colitis. *J Invest Dermatol.* 1953; **20**: 353–6.
28 Finkel S I, Janowitz H D. Trauma and the pyoderma gangrenosum of inflammatory bowel disease. *Gut.* 1981; **22**: 410–2.
29 Gelernt I M, Kreel I. Pyoderma gangrenosum in ulcerative colitis: prevention of the gangrenous component. *Mount Sinai J Med.* 1976; **43**: 467–70.
30 Basler R S W, Dubin H V. Ulcerative colitis and the skin. *Arch Dermatol.* 1976; **112**: 531–4.
31 Lazarus G S, Goldsmith L A, Rocklin R E, *et al.*. Pyoderma gangrenosum, altered delayed hypersensitivity and polyarthritis. *Arch Dermatol.* 1972; **105**: 46–51.
32 Solomon D H, Chopra I J, Chopra U, Smith F J. Identification of sub-groups of euthyroid Graves' ophthalmopathy. *N Engl J Med.* 1977; **296**: 181–6.

13

Mycobacterial disease of the gut

Michael J.G. Farthing

Introduction

Mycobacterial disease is one of the most common human infections. Human pathogenic mycobacteria include *M tuberculosis* and *M bovis* both of which produce the systemic infection of tuberculosis (TB) and *M leprae*, which causes leprosy (Table 13.1). It is estimated that for *Mycobacterium tuberculosis* alone there are approximately 8 million new infections each year with an annual mortality of approximately 3 million.

Some of the atypical mycobacteria (slower growing and generally resistant to standard anti-TB drugs) can also cause disease in humans although in general they are opportunistic pathogens (Table 13.1). The *M avium-intracellulare* group of organisms produce a tuberculosis-like illness as an opportunistic infection in immunocompromised hosts. Disseminated *M avium* complex infection is increasingly recognized in AIDS patients, with more than 20 per cent eventually acquiring the infection. *M paratuberculosis* is a related organism that is responsible for the granulomatous enteritis of ruminants (Johne's disease) and until recently was thought not to produce disease in humans. However, evidence now exists

Table 13.1 Mycobacteria and human disease

	Disease
Pathogenic mycobacteria	
M. tuberculosis	Tuberculosis
M. bovis	Tuberculosis
M. leprae	Leprosy
Potentially pathogenic mycobacteria*	
M. avium complex	Tuberculosis-like
M. kansasii	Tuberculosis-like
M. paratuberculosis	? Crohn's disease
M. fortuitum complex	Wound infections
M. marinum	Skin lesions

*Atypical mycobacteria.

to suggest that this organism may be involved in the aetiopathogenesis of Crohn's disease.

Mycobacterium tuberculosis

Abdominal tuberculosis was recognized by Hippocrates in the 4th Century BC when he realized that 'phthisical persons die if diarrhoea sets in'. Louis XIII died in 1643, probably of tuberculosis with intestinal involvement, because his autopsy showed intestinal ulceration, perforation and peritonitis in addition to pulmonary cavities.[1] John Hunter recognized the importance of abdominal tuberculosis when writing in 1786 he stated 'I have seen the whole intestine adhering to the peritoneum seemly from a scrofulous case and having scrofulous tumours and suppurations in them also; the symptoms were tightness of the belly without pain, and costiveness; at other times purging'.[2] It was not until 1882 that Robert Koch identified the tubercle bacillus as the cause of human tuberculosis.

Epidemiology

Abdominal tuberculosis is found worldwide although prevalence rates are still highest in the developing world. Due to immigration, however, abdominal tuberculosis is increasing in some developed countries.[2–10] In the indigenous population of developed communities, the incidence of abdominal tuberculosis is extremely low and in one city in the UK it has been estimated at 0.43/100 000 whereas the Asian immigrant population in the same area has an incidence of 35.7/100 000.[4] It is now well established that immigrants from developing communities are commonly infected with *M tuberculosis* and less commonly with *M bovis*.[3] In five studies from the UK between 1980–86 involving 395 patients, the average age at presentation was 33 years (range 3–71 years), the sex ratio was equal and 82–87 per cent of the patients in these series were immigrants.[2–6] However, it should not be forgotten that *M tuberculosis* is endemic in the UK and that the indigenous population are at risk of developing abdominal tuberculosis.

In the past *M bovis* has generally been associated with abdominal tuberculosis infection being transmitted in the milk of infected cows. However, in many parts of the world this infection has been controlled both by ensuring that dairy herds are tuberculosis-free and by pasteurization of milk. In the Blackburn study of 109 patients only 3 per cent were infected with *M bovis*, the remaining have *M tuberculosis*.[4]

For the past ten years there has been an increasing number of cases of abdominal tuberculosis in patients with AIDS. During an 18-month period in a major New York City Hospital, of 21 patients with abdominal mycobacterial infection, 14 (67 per cent) had AIDS.[11] Ileocaecal and peritoneal involvement were the most common manifestations. Overall the clinical presentation was similar to that of non-AIDS patients although surgery was required in ten of the 14 AIDS patients; procedures included drainage of three psoas abscesses and a paracolic abscess, resection of ileal strictures, a diagnostic laparotomy and laparotomy for perforation.

Pathogenesis

Tubercle bacilli probably infect the abdominal cavity and its organs either directly by being swallowed or by haematogenous spread during a primary pulmonary infection.[10,12] It is thought that tubercle bacilli cross the epithelium and enter Peyer's patches most commonly in the distal ileum and then spread into mesenteric lymph nodes to cause hyperplasia and subsequent caseation and necrosis. Retrograde spread of bacilli may occur from these lymph nodes to the bowel wall where it is thought that an inflammatory endarteritis leads to transmural inflammation, ulceration and sometimes perforation. Experimental infections in dogs with Thiry-Vella fistulae confirm that bacilli can transmigrate the intestinal wall and remain viable in venous blood.[10] It is also thought that haematogenous spread of bacilli can also occur from a primary pulmonary infection possibly accounting for disseminated peritoneal infections. Pathogenic mycobacteria do not produce any known classic virulence factors such as toxins, proteolytic enzymes or haemolysin but their pathogenicity appears to rest solely on their capacity to multiply and survive within macrophages. The presence of bacilli in tissue initiates an inflammatory cellular immune response leading to a local necrotizing allergic reaction that is very characteristic of the destructive process of progressive chronic mycobacterial infection.

Clinical features

Abdominal tuberculosis has three major categories of clinical presentation, (i) gastrointestinal disease, (ii) mesenteric lymphadenopathy and (iii) peritonitis.

Gastrointestinal disease

Ileocaecal disease

This form of abdominal tuberculosis is found both in the tropics and in immigrants to industrialized communities. In the UK studies it accounted for 40–60 per cent of patients with abdominal tuberculosis and in patients with intestinal disease it is the site of predeliction.[2–6] Pain is the prominent feature of infection in this region of the intestinal tract. It may be either obstructive in character, due to stricturing of the terminal ileum, in which case it is generally experienced in the central abdomen, or localized to the right iliac fossa as a result of transmural focal inflammation. A mass may be palpable and there is often fever, diarrhoea and general malaise. Perforation, although unusual, can occur producing generalized abdominal pain due to peritonitis.[13,14] Of 42 patients with ileocaecal disease reported by Palmer *et al.*,[5] all had abdominal pain, 43 per cent had an abdominal mass, 52 per cent had weight loss and 60 per cent were febrile.

Colonic and anorectal disease

Infection is sometimes limited to the colon producing colicky lower abdominal pain, altered bowel habit and fever.[15,16] Stricture formation is a common complication. Tuberculosis may also involve the anal canal where it initially produces a painless ulcer, indistinguishable on inspection from a simple anal fissure[17,18] The ulcer, however, tends to enlarge and becomes indurated with undermined edges. Anorectal disease is sometimes complicated by fistula formation and abscesses that may involve the ischiorectal space. Shukla *et al.*[19] found that 16 per cent of fistulae examined in Indian patients were tuberculous in origin. A fistula should be considered tuberculous if the opening on to the skin is ragged, when induration is mild or absent and if the discharge is watery.

Oesophageal and gastroduodenal disease

The most common presentation of oesophageal involvement with *M tuberculosis* is dysphagia.[20,21] This may be due either to extrinsic compression from enlarged mediastinal nodes or from a mass lesion within the oesophagus itself. Discrete tuberculous ulcers may also be found in the oesophagus (Fig. 13.1) and infection in this region may be complicated by broncho-oesophageal fistula. It has been suggested that oesophageal involvement is most likely to be due to retrograde spread of bacilli from mediastinal lymph nodes. Gastroduodenal involvement also occurs[2,21,22] with ulceration in both the stomach and duodenum resembling typical peptic ulcers; the presence of an inflammatory mass in the pre-pyloric region may mimic Crohn's disease or neoplasia.

The complications of gastrointestinal tuberculous disease include intestinal obstruction involving both the ileum and colon. Perforation may occur leading either to a diffuse peritonitis or a localized abscess within the abdominal cavity or in the perianal region. Fistulation occurs both internally and externally, and occasionally a tuberculous ulcer may severely haemorrhage.

Mesenteric lymphadenopathy

This is the most common variety of abdominal tuberculosis found in the tropics and may be associated with lymphadenopathy in other locations although a pulmonary lesion is notable by its absence. The illness begins insidiously with weight loss, intermittent low grade fever and general malaise. As the disease progresses, abdominal swelling occurs due both to accumulation of fluid within the abdominal cavity and also the often massively enlarged lymph nodes. If the disease is allowed to progress, anaemia, hypo-albuminaemia and peripheral oedema, often with lymphoedema become evident. Massive caseation of mesenteric lymph nodes occurs, and a major complication of this form of abdominal tuberculosis is node rupture with dissemination of bacilli throughout the abdominal cavity causing tuberculous peritonitis with multiple tubercles 'peppering' the peritoneal surface. This form of the disease may mimic lymphoma.

Tuberculous peritonitis

Peritoneal involvement probably accounts for 25–30 per cent of abdominal tuberculosis in the tropics and a similar or even higher proportion of

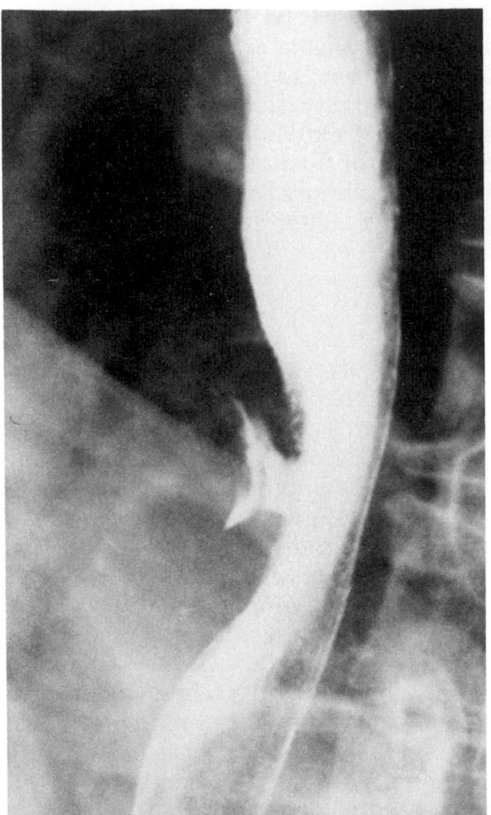

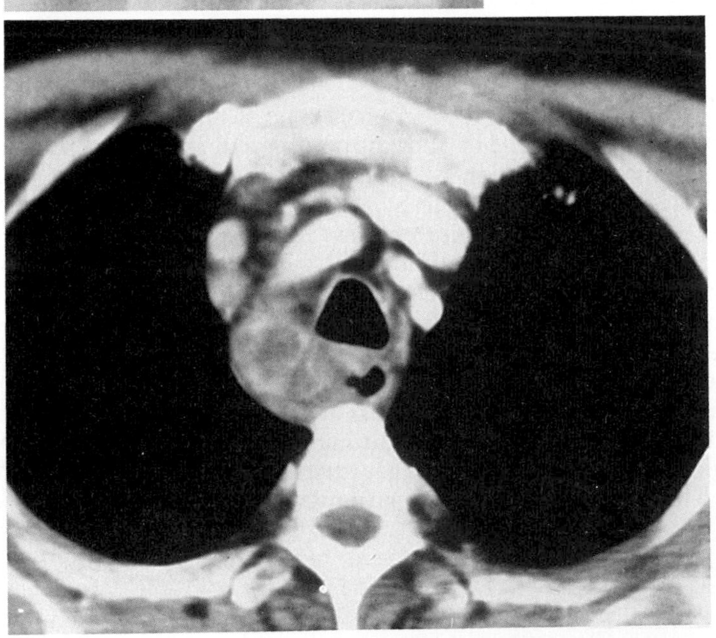

Fig. 13.1 Barium swallow examination showing penetrating tuberculous ulcer in the oesophagus. Computed tomography scan shows associated enlarged mediastinal lymph nodes.

disease in immigrant patients in developing communities[6,21,23] Peritoneal tuberculosis generally has an insidious onset usually involving several months duration. Symptoms include fever, weight loss, anorexia and lethargy. In the so-called 'wet' form of the disease there is progressive ascites and in the more advanced 'dry' form, bowel loops and omentum are matted together giving rise to abdominal pain and tenderness, moderate distension, a mass in approximately 25 per cent and a doughy feel to the abdomen.[23] Such patients may occasionally present with obstructive features secondary to tuberculous adhesions. Peritonitis may begin abruptly following bowel perforation or massive rupture of caseating abdominal lymph nodes, but is most commonly due to reactivation of latent peritoneal disease following haematogenous spread from a primary focus.[24]

Diagnosis

In all cases of suspected abdominal tuberculosis, one should attempt to achieve a microbiological diagnosis, by the identification of *M tuberculosis* or *M bovis* in tissue or secretions or the presence of caseating granulomata in tissues. In the developing world where abdominal tuberculosis is relatively common, treatment is often given on the basis of clinical history and physical findings alone, but this approach has the disadvantage of offering potentially toxic treatment for prolonged periods without knowledge of the drug sensitivity profile of the organism responsible. In the industrialized world there are added problems in that gastrointestinal tuberculosis may be confused with other conditions, such as Crohn's disease, which, although uncommon in the developing world, is recognized increasingly in Asian and other immigrants.

Although a rapid erythrocyte sedimentation rate is present in more than 80 per cent of patients, the white blood cell count increased in 30–40 per cent and a normochromic, normocytic anaemia present in more than 50 per cent, these tests are poor diagnostic discriminators.[2–6] Tuberculin testing either by the Mantoux or Heaf test are also of limited diagnostic value with positivity varying from 30–100 per cent in different series.[22,25,26] Patients with abdominal tuberculosis generally have weakly positive tests rather than the strongly positive result normally seen in patients with active pulmonary tuberculosis.

Gastrointestinal tuberculosis may be diagnosed at the laparotomy when presenting acutely with intestinal obstruction or perforation. Biopsy material should be sent for culture and histopathological examination if acid-fast bacilli are detected and the presence of caseating granulomata confirmed. Barium-contrast radiology most commonly reveals abnormalities in the ileocaecal region with ulceration and stricturing of the terminal ileum, obliteration of the ileocaecal angle and shortening of the ascending colon and caecum[27–29] (Fig. 13.2). Barium enema examination may reveal colonic strictures (Fig. 13.3). However, these radiological appearances are not specific for tuberculosis and cannot be reliably differentiated from Crohn's disease. Similar appearances may less commonly be seen with

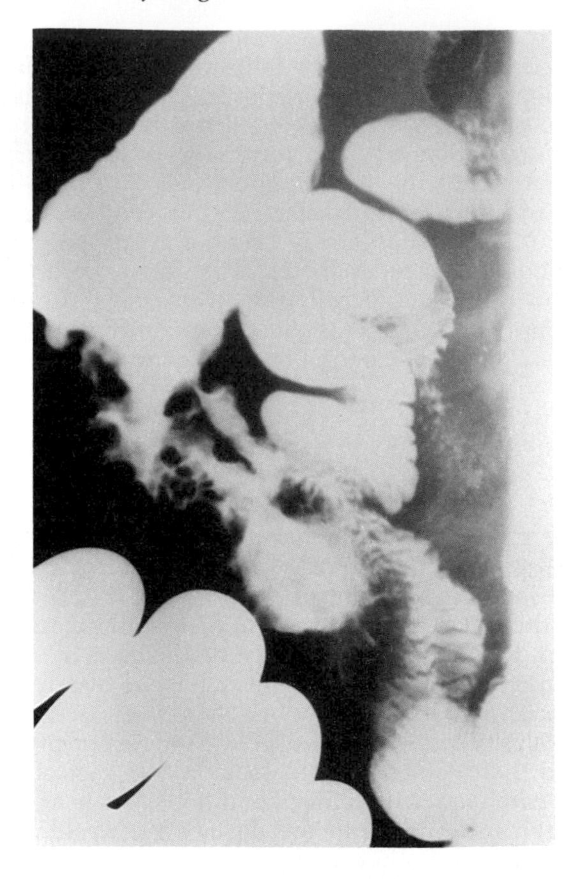

Fig. 13.2 Barium follow-through examination showing ileocaecal tuberculosis with contraction of the caecum, shortening of the ascending colon and ileal ulceration.

lymphosarcoma, radiation ileitis, carcinoid tumours and actinomycosis. Endoscopic examination of the gastrointestinal tract can have an important part to play because of the obvious advantage of being able to obtain tissue for culture and microscopic examination.[30–34] The macroscopic appearances may suggest tuberculosis but are rarely diagnostic.[33] In the ileum and colon typical lesions include mucosal nodules, superficial and deep ulceration (Fig. 13.4) particularly in the caecum and a deformed ileocaecal valve. Aphthoid ulcers, typical of Crohn's disease, are also found. Like Crohn's disease the changes are patchy with normal mucosa intervening between areas of inflammation or ulceration. Similar appearances may be found in the oesophagus and other parts of the upper gastrointestinal tract. Endoscopic biopsies in some series have been relatively unrewarding. However, when biopsies are taken with forceps that take deeper biopsies

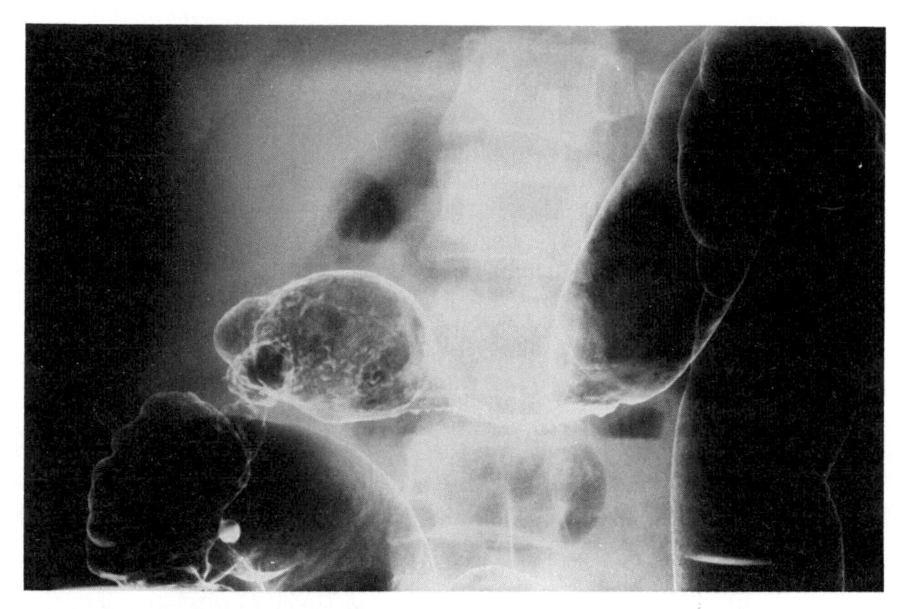

Fig. 13.3 Double contrast barium enema examination showing tuberculous strictures in the transverse colon.

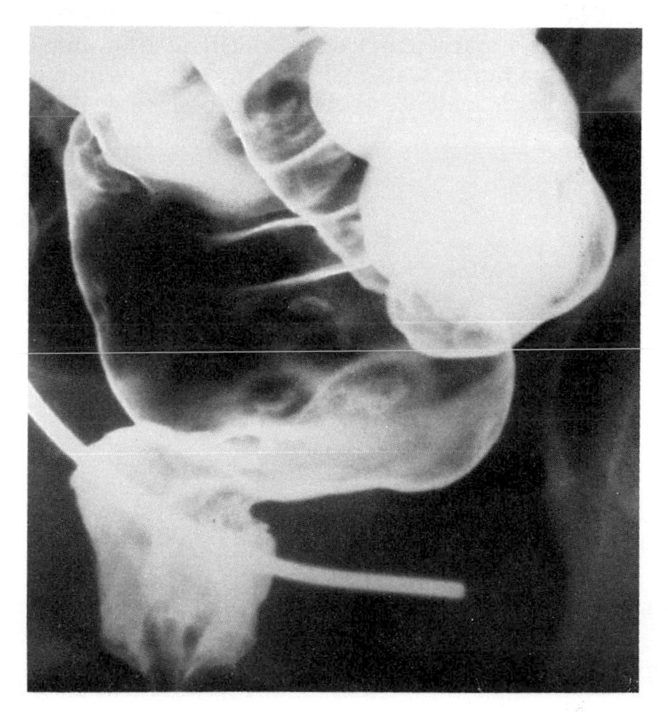

Fig. 13.4 Double contract barium enema examination showing deep colonic tuberculous ulcers.

ensuring the presence of submucosa, the diagnostic rate can be substantially increased.[33] Searching for mycobacteria in faeces in patients with suspected gastrointestinal tuberculosis is unrewarding.

In patients suspected of having *tuberculous peritonitis*, abdominal ultrasound is the first-line investigation. This may detect bowel wall thickening,[35] detect subclinical ascites and mesenteric lymph node enlargement. In addition, it may identify involvement of both the visceral and parietal peritoneum. When ascites is present this should be aspirated; the fluid is usually clear with a high protein concentration and rarely blood stained or chylous. Direct smears of the ascitic fluid for acid-fast bacilli have a low yield rate whereas culture is positive in 34–67 per cent of patients in several series.[23] When examination of ascitic fluid fails to give a positive diagnosis or when it is impossible to aspirate fluid, many investigators favour laparoscopy and guided biopsy.[36,37] The presence of multiple tubercles disseminated throughout the peritoneal cavity strongly supports the diagnosis and although biopsy is not always positive for acid-fast bacilli, diagnosis can be increased to 85 per cent by culture of biopsy tissues.[38] Although blind peritoneal biopsy has had its successes in the past,[39] it is almost certainly safer to use either the laparoscopic or endoscopic route to obtain biopsy material. If all of these approaches fail, diagnostic laparotomy may be required although in very ill patients with advanced disease it may well be advisable to commence anti-tuberculous chemotherapy before a microbiological diagnosis is confirmed.

Enlargement of *mesenteric lymph nodes* is sometimes apparent on a plain abdominal radiograph, particularly when calcified. Abdominal ultrasound or a computed tomography scan will confirm the presence of mesenteric lymphadenopathy with evidence of central necrosis (Fig. 13.5). Fine needle

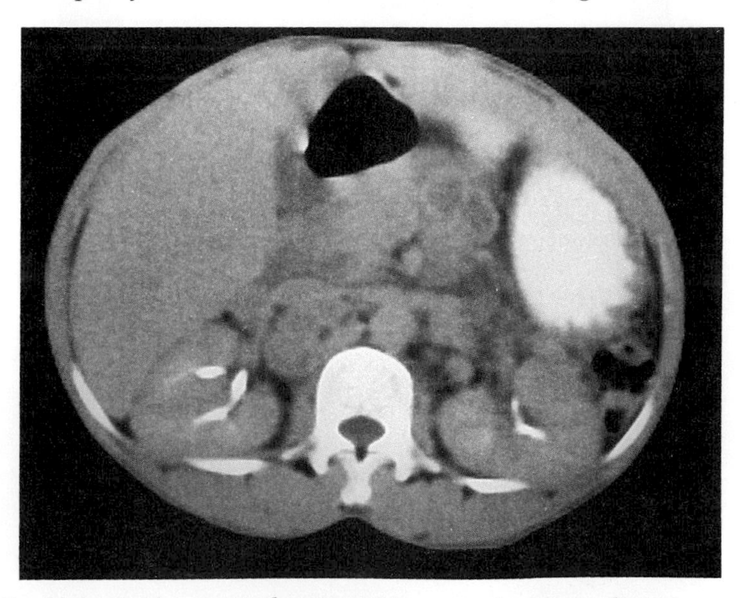

Fig. 13.5 Abdominal computed tomography scan showing enlarged mesenteric lymph nodes with central caseation due to tuberculosis.

aspiration, guided by either of these imaging techniques is feasible, although the chance of making a microbiological diagnosis by this route is probably not high. Other causes of massive mesenteric lymphadenopathy, such as lymphoma, can be confused with tuberculosis particularly when nodes enlarge rapidly and when associated with central necrosis. Laparotomy and excision biopsy of the lymph nodes should be considered if the diagnosis can not be achieved by less invasive procedures. An approach to the diagnosis of abdominal tuberculosis is shown in Fig. 13.6.

Most patients with abdominal tuberculosis do not have active pulmonary disease and less than half will have an abnormal chest radiograph.[2, 5, 6, 22] When the clinical presentation strongly suggests abdominal tuberculosis and focal signs are absent, liver biopsy may reveal caseating granulomata,[40] particularly if serum alkaline phosphatase is elevated.

Although immunological tests have not been helpful in the diagnosis of tuberculosis, a modified serological competition assay directed towards the 30kDa *M tuberculosis* antigen detected 73 per cent of cases of extrapulmonary tuberculosis and 70 per cent of smear-negative pulmonary tuberculosis.[41] Detection of *M tuberculosis* in tissues and body fluids is now possible using the polymerase chain reaction. This technique has been successfully used on cerebrospinal fluid using a 240 bp region from the gene coding the MPB64 protein.[42] This approach should be applicable to ascitic fluid, mesenteric lymph node aspirates and intestinal biopsy specimens.

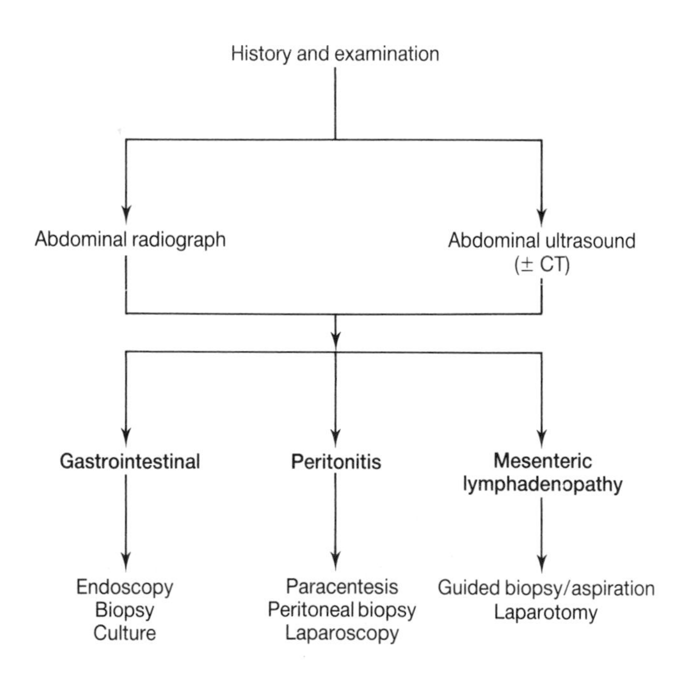

Fig. 13.6 An approach to the diagnosis of abdominal tuberculosis.

Treatment

The current recommendations of the British Thoracic Society for the treatment of extrapulmonary tuberculosis is that daily isoniazide (330 mg) and rifampicin (450–600 mg) should be given for 6 months, with pyrazinamide (20–30 mg/kg daily, maximum 3 g daily) included for the first 2 months.[43] A fourth drug such as streptomycin or ethambutol should be added initially if drug resistance is suspected, particularly in patients who may have imported the disease from a developing country. Major adverse reactions are not common but liver biochemistry should be assessed before treatment starts. Aminotransferase concentrations often increase transiently during treatment but the drug should be continued unless symptoms of hepatitis or jaundice develop. If hepatitis develops all drugs should be withdrawn until liver biochemistry returns to pretreatment levels. Treatment can then usually be resumed without recurrence of the hepatitis. Renal function should be checked before starting streptomycin or ethambutol and these drugs should be avoided in patients with impaired renal function. Visual acuity should be checked before starting ethambutol.

Abdominal disease and atypical mycobacteria

Atypical mycobacteria are widely disseminated in the environment and until recently have not been considered major human pathogens. However, they are opportunistic organisms some of which are now recognized to cause important infections in AIDS patients and some evidence exists to suggest that *M paratuberculosis* may be involved in the pathogenesis of Crohn's disease in some patients.

Mycobacterium avium complex infection in AIDS

M avium and *M intracellulare* have little pathogenic potential in the normal human host although *M avium* complex is an uncommon cause of pneumonia in individuals with chronic lung disease.[44] However, during the 1980s there was a progressive increase in the number of AIDS patients with *M avium* complex infection and it has been estimated that more than 20 per cent of patients with AIDS will become infected.[45] Infection usually follows the AIDS diagnosis and most commonly occurs when the CD4+ count is less than 100. Unlike *M tuberculosis* in AIDS, which is thought to be a reactivation of a primary infection, *M avium* complex infection is thought to be newly acquired.[45] The organism is widely dispersed in the environment and can be acquired from water and food sources and possibly by aerosol spread. The major difference between colonization of an immunocompetent host and an individual with AIDS is that in the former colonization remains localized to the respiratory or gastrointestinal tract, but in a person with HIV infection may become disseminated. Recovery of organisms can be made from almost any tissue in the body.

Clinical features

Persistent fever and weight loss is the most common presentation.[45] However, chronic diarrhoea with malabsorption, abdominal pain and extrahepatic biliary obstruction also occur.[46–48] Severe anaemia is characteristic of the infection and may require regular transfusion. Rare manifestations include localized pneumonia, arthritis, skin lesions and endophthalmitis.

Diagnosis

Organisms can be cultured from blood in 75–100 per cent of cases but can also be obtained from bone marrow, lymph nodes and liver.[45] Tissue specimens contain large numbers of acid-fast bacilli and also may contain small numbers of ill-defined granulomas. Involvement of the small intestine produces a microscopic picture similar to Whipple's disease.[46,48] The mucosal macrophages and regional lymph nodes are filled with foamy macrophages that are full of acid-fast bacilli.

Treatment

Atypical mycobacteria are more resistent to anti-tuberculous chemotherapy and thus early studies with two or three drugs did not generally produce favourable results.[45,49] However, using four or five drug regimens with agents such ciprofloxacin, clofazamine, ethambutol, and rifampacin, with the addition of amikacin if there is no response to the above drugs in 4–6 weeks, can substantially improve symptoms although current evidence suggests that it will not eradicate infection.[45] All of these drugs are toxic and the relative risks and benefits need to be assessed in individual patients before the decision to treat is made. New drugs may, however, be available within the foreseeable future.

M paratuberculosis and Crohn's disease

M paratuberculosis is known to cause the chronic enteritis of cattle known as Johne's disease. This is an important disease of cattle, is found worldwide and is a major cause of morbidity and mortality.

Dalziel, a Glasgow surgeon recognized a similar disease in humans, and knowing about Johne's disease considered that this might be a related problem, but was unable to detect acid-fact bacilli.[50] Since Crohn *et al.*[51] formally described regional ileitis in 1932 there has been a continuing search for *Mycobacterium* species in Crohn's disease. A major step forward occurred in 1984 when Chiodini *et al.* isolated *M paratuberculosis* from 4 of 26 patients with Crohn's disease after 9–18 months culture.[52] Other groups have also been able to culture *M paratuberculosis* from Crohn's disease patients and a variety of other acid-fast mycobacteria including *M avium-intracellulare, M cheloni-fortuitum* and *M kansasii*. Combining all studies mycobacteria and acid-fast bacilli have been cultured from the tissues of approximately 20 per cent of the Crohn's disease cases and approximately 10 per cent of the controls. However, these mycobacteria can also be cultured from approximately 20 per cent of patients with ulcerative colitis.[53] A major problem with this experimental approach is that these mycobacteria grow extremely slowly and the question as to whether they are truly present within the gut tissues or are merely luminal contaminants or opportunistic organisms in damaged tissue has as yet not been resolved.

An alternative approach has been to search for mycobacterial DNA, specifically *M paratuberculosis* DNA, within tissue samples by *in situ* hybridization and the polymerase chain reaction. Using one of the original Chiodini isolates, genomic DNA was digested and cloned into a plasmid vector. Using a number of clones as probes, it was possible to distinguish between mycobacterial species by restriction fragment length polymorphism.[54,55] A DNA insertional element IS900 from one of the original clones pMB22 has been sequenced and is highly specific for *M paratuberculosis*.[53] Polymerase chain reaction (PCR) amplification of this region is being used in an attempt to detect *M paratuberculosis* DNA in tissue extracts. Preliminary work suggests that approximately 65 per cent of patients with Crohn's disease have positive tests but data on appropriate inflammatory bowel disease controls are as yet not available.[56] Preliminary data from other groups, however, are inconsistent, some confirming these findings, and others not.

Attempts at serological detection of mycobacterial antigens in Crohn's disease have proved variable and generally lack specificity. Thayer *et al.* initially reported that 23 per cent of Crohn's disease cases had increased antibody titres to a crude fraction of *M paratuberculosis*[57] but subsequent studies using purified antigen preparations have failed to detect differences between Crohn's disease patients and controls.[58, 59]

Even if it can be demonstrated that there is an association between *M paratuberculosis* and Crohn's disease it will be extremely difficult to argue that the bacillus is aetiologically important and not merely an opportunistic infection in damaged tissue. The fact that these organisms have been detected in patients with ulcerative colitis would support this view. Furthermore, the fact that it is so difficult to isolate the organisms by culture and that *in situ* hybridization without PCR amplification fails to detect *M paratuberculosis* DNA suggests that even if the organism is present it is only there in very small numbers. Thus, the initial trigger could have been *M paratuberculosis*, but active ongoing infection seems to be of questionable pathogenetic significance, particularly as the disease commonly responds well to immunosuppressive therapy.

Anti-mycobacterial therapy in Crohn's disease

An alternative approach to determine the relevance of *M paratuberculosis* and other atypical mycobacteria in Crohn's disease is to attempt to treat the disease with anti-tuberculous chemotherapy. Unfortunately these organisms are relatively insensitive to currently available anti-tuberculous drugs and thus a negative result would not necessarily exclude their role in aetiopathogenesis. Similarly, a positive result would not necessarily implicate mycobacteria because many of the drug regimens used are not monospecific; rifampicin, for example, is an excellent antimicrobial drug with a broad antibacterial spectrum. Early studies giving one or two drugs produced uniformly disappointing results.[60] However, regimens with triple or quadruple therapy in uncontrolled studies have produced potentially encouraging results in selected patients.[61,62] However, these drugs are toxic and until the results of double-blind, controlled clinical trials are known, widespread use cannot be recommended.

Acknowledgements

MJGF gratefully acknowledges financial support by the Wellcome Trust, The Joint Research Board of St Bartholomew's Hospital, The Royal Society, The Mason Medical Foundation and The British Digestive Foundation.

References

1 Goldfischer S, Janis M. A 42-year old king with a cavitatory pulmonary lesion and intestinal perforation. *Bull NY Acad Med*. 1981; **57**: 139–43.
2 Addison N V. Abdominal tuberculosis – a disease revived. *Ann R Coll Surg Engl*. 1983; **65**: 105–111.
3 Lambrianides A L, Ackroyd N, Shorey B A. Abdominal tuberculosis. *Br J Surg*. 1980; **67**: 887–9.
4 Klimach O E, Ormerod L P. Gastrointestinal tuberculosis: A retrospective review of 109 cases in a district general hospital. *Q J Med*. 1985; **56**: 569–78.
5 Palmer K R, Patil D H, Basran G S, *et al*.. Abdominal tuberculosis in urban Britain – a common disease. *Gut*. 1985; **26**: 1296–305.
6 Wells A D, Northover J M A, Howard E R. Abdominal tuberculosis: still a problem today. *J R Soc Med*. 1986; **79**: 149–53.
7 Schofield P F. Abdominal tuberculosis. *Gut*. 1985; **26**: 1275–8.
8 Cook G C. Tuberculosis – certainly not a disease of the past! *Q J Med*. 1985; **56**: 519–21.
9 Panton O N M, Sharp R, English R A, Atkinson K G. Gastrointestinal tuberculosis. The great mimic still at large. *Dis Colon Rectum*. 1985; **28**: 446–50.
10 Wald A. Enteric tuberculosis: Literature review. *Mt Sinai J Med*. 1987; **54**: 443–9.
11 Rosengart T K, Coppa G F. Abdominal mycobacterial infections in immunocompromised patients. *Am J Surg*. 1990; **159**: 125–31.
12 Vanderpool D M, O'Leary J P. Primary tuberculous enteritis. *Surg Gynecol Obstet*. 1988; **167**: 167–73.
13 Porter J M, Showe R J, Silver D. Tuberculous enteritis with perforation and abscess formation in childhood. *Surgery*. 1972; **71**: 254–7.
14 Gilinsky N H, Marks I N, Kottler R E, Price S K. Abdominal tuberculosis: a 10-year review. *S Afr Med J*. 1983; **64**: 849–57.
15 Ahjua A K, Gaiha M, Sachdev S. Tubercular colitis simulating ulcerative colitis. *J Assoc Physicians India*. 1976; **24**: 617–9.
16 Balikian J P, Uthman S M, Kabakian H A. Tuberculous colitis. *Am J Proctol*. 1977; **28**: 75–9.
17 Gupta A S, Sharma V P, Rathi G L. Anorectal tuberculosis simulating carcinoma. *Am J Proctol*. 1976; **27**: 33–8.

18 Goyal S C, Singh K P, Sabharwal B D, Bhandari Y P. Granulomatous lesions of rectum. *J Indian Med Assoc*. 1977; **69**: 16–17.
19 Shukla H S, Gupta S C, Singh G, Singh P A. Tubercular fistula-in-ano. *Br J Surg*. 1988; **75**: 38–9.
20 McNamara M, Williams C E, Brown T S, Gopichandra T D. Tuberculosis affecting the oesophagus. *Clin Radiol*. 1987; **38**: 419–22.
21 Yasawy M I, Al Karawi M A, Mohamed A E. Alimentary tract tuberculosis. A continuing challenge to gastroenterologists – report of 55 cases. *J Gastroenterol Hepatol*. 1987; **2**: 137–47.
22 Mandal B K, Schofield P F. Abdominal tuberculosis in Britain. *Practitioner*. 1976; **216**: 683–9.
23 Bastani B, Shariatzdeh M R, Dehdasti F. Tuberculous peritonitis – report of 30 cases and review of the literature. *Q J Med*. 1985; **56**: 549–57.
24 Nice C M. Pathogenesis of tuberculosis. *Dis Chest*. 1950; **17**: 550–60.
25 Wales J M, Mumtaz H, Macleod W M. Gastrointestinal tuberculosis. *Br J Dis Chest*. 1976; **70**: 39–57.
26 Findlay J M, Addision N V, Stevenson B K, Mirza Z A. Tuberculosis of the gastrointestinal tract in Bradford 1967–1977. *J R Soc Med*. 1979; **72**: 587–90.
27 Anscombe A R, Keddie N C, Schofield P F. Caecal tuberculosis. *Gut*. 1967; **8**: 337–43.
28 Prakash A, Sharma L K, Koshal A, Poddar P K. Ileocaecal tuberculosis. *Aust NZ J Surg*. 1975; **45**: 371–5.
29 Lanzieri L F, Keller R J. Primary tuberculosis of the ileum: roentgen features. *Mt Sinai J Med*. 1980; **47**: 596–9.
30 Franklin G O, Mohaptra M, Perillo R P. Colonic tuberculosis diagnosed by colonoscopic biopsy. *Gastroenterology*. 1979; **76**: 362–4.
31 Breiter J R, Hajjar J J. Segmental tuberculosis of the colon diagnosed by colonoscopy. *Am J Gastroenterol*. 1981; **76**: 369–73.
32 Radhakrishnan S, Alnakib B, Shaikh H, Menon N K. The value of colonoscopy in schistosomal, tuberculous and amoebic colitis. *Dis Colon Rectum*. 1986; **29**: 891–5.
33 Kalvaria I, Kottler R E, Marks I N. The role of colonoscopy in the diagnosis of tuberculosis. *J Clin Gastroenterol*. 1988; **10**: 516–23.
34 Morgante P E, Gandara M A, Sterle E. The endoscopic diagnosis of colonic tuberculosis. *Gastrointest Endosc*. 1989; **35**: 115–8.
35 Bluth E I, McVay L V, Gathright J B. Ultrasonic characteristics of ileal tuberculosis. *Dis Colon Rectum*. 1985; **28**: 613–4.
36 Wolfe J H N, Behn A R, Jackson B T. Tuberculous peritonitis and the role of diagnostic laparoscopy. *Lancet*. 1979; **i**: 852–3.
37 Menzies R I, Fitzgerald J M, Mulpeter K. Laparoscopic diagnosis of ascites in Lesotho. *Br Med J*. 1985; **291**: 473–5.
38 Singh M M, Bhargava A N, Jain K P. Tuberculous peritonitis – an evaluation of pathogenic mechanisms, diagnostic procedures and therapeutic measures. *N Engl J Med*. 1969; **281**: 1091–4.
39 Das P, Shukla H S. Clinical diagnosis of abdominal tuberculosis. *Br J Surg*. 1976; **63**: 941–6.
40 Gambhir M S, Goyal S K, Rewat M L. Hepatic involvement in abdominal tuberculosis: A clinical, biochemical and histopathological study. *J Assoc Physicians India*. 1972; **20**: 854–74.
41 Wilkins E G L, Ivanyi J. Potential value of serology for diagnosis of extrapulmonary tuberculosis. *Lancet*. 1990; **336**: 641–4.
42 Shankar P, Manjunath N, Mohan K K, *et al.*. Rapid diagnosis of tuberculous meningitis by polymerase chain reaction. *Lancet*. 1991; **337**: 5–7.

43 Ormerod L P. Chemotherapy and management of tuberculosis in the United Kingdom: recommendations of the Joint Tuberculosis Committee of the British Thoracic Society. *Thorax*. 1990; **45**: 403–408.
44 Wolinsky E. Non-tuberculous mycobacteria and associated diseases. *Am Rev Resp Dis*. 1979; **119**: 107–59.
45 Horsburgh C R. *Mycobacterium avium* complex infection in the acquired immunodefiency syndrome. *N Engl J Med*. 1991; **324**: 1332–8.
46 Strom R L, Gruninger R P. AIDS with *Mycobacterium avium-intracellulare* lesions resembling those of Whipple's disease. *N Engl J Med*. 1983; **309**: 1323–4.
47 Roth R I, Owen R L, Keren D F, Volberding P A. Intestinal infection with *Mycobacterium avium* in acquired immune deficiency syndrome (AIDS). *Dig Dis Sci*. 1985; **30**: 497–504.
48 Vincent M E, Robbins A H. *Mycobacterium avium-intracellulare* complex enteritis: Pseudo-Whipple disease in AIDS. *Am J Roenterol*. 1985; **144**: 921–2.
49 Glatt A E, Chirgwin K, Landesman S H. Treatment of infections associated with human immunodeficiency virus. *N Engl J Med*. 1988; **318**: 1439–48.
50 Dalziel T K. Chronic intestinal enteritis. *Br Med J*. 1913: **ii**: 1068–70.
51 Crohn B B, Ginzburg L, Oppenheimer G D. Regional ileitis, a pathologic and clinical entity. *J Am Med Assoc*. 1932; **99**: 1323–9.
52 Chiodini R J, Van Kruiningen J H, Thayer W R, *et al.*. Possible role of mycobacteria in inflammatory bowel disease. *Dig Dis Sci*. 1984; 1073–9.
53 Hermon-Taylor J, Moss M, Tizard M, *et al.*. Molecular biology of Crohn's disease mycobacteria. *Clin Gastroenterol*. 1990; **4**: 23–42.
54 McFadden J J, Butcher P D, Chiodini R, Hermon-Taylor J. Crohn's disease-isolated mycobacteria are identical to *Mycobacterium paratuberculosis* as determined by DNA probes that distinguish between mycobacterial species. *J Clin Microbiol*. 1987; **25**: 796–801.
55 McFadden J J, Butcher P D, Thompson J, *et al.*. The use of DNA probes identifying restriction fragment length polymorphism to examine the *Mycobacterium avium* complex. *Mol Microbiol*. 1987; **1**: 283–91.
56 Sanderson J D, Moss M, Malik Z, *et al.*. Polymerase chain reaction detects *Mycobacterium paratuberculosis* in Crohn's disease tissue extracts. *Gut*. 1991; **32**: A572.
57 Thayer W R, Coutu J A, Chiodini R J, *et al.*. Possible role of mycobacteria in inflammatory bowel disease. II. Mycobacterial antibodies in Crohn's disease. *Dig Dis Sci*. 1984; **29**: 1080–5.
58 Kobayashi K, Brown W R, Brennan P J, Blaser M J. Serum antibodies to mycobacterial antigens in active Crohn's disease. *Gastroenterology*. 1988; **94**: 1404–11.
60 Shaffer J L, Hughes S, Linaker B D, *et al.*. Controlled trial of rifampicin and ethambutol in Crohn's disease. *Gut*. 1984; **25**: 203–205.
61 Warren J B, Rees H C, Cox T M. Remission of Crohn's disease with tuberculosis chemotherapy. *N Eng J Med*. 1986; **314**: 182.
62 Hampson S J, Parker M C, Saverymuttu S H, *et al.*. Quadruple anti-mycobacterial chemotherapy in Crohn's disease: results at 9 months of a pilot study in 20 patients. *Aliment Pharmacol Therapeut*. 1989; **3**: 343–52.

Index